Navigation: A Very Short Introduction

VERY SHORT INTRODUCTIONS are for anyone wanting a stimulating and accessible way into a new subject. They are written by experts, and have been translated into more than 45 different languages.

The series began in 1995, and now covers a wide variety of topics in every discipline. The VSI library now contains over 500 volumes—a Very Short Introduction to everything from Psychology and Philosophy of Science to American History and Relativity—and continues to grow in every subject area.

Very Short Introductions available now:

ACCOUNTING Christopher Nobes
ADOLESCENCE Peter K. Smith
ADVERTISING Winston Fletcher
AFRICAN AMERICAN RELIGION
  Eddie S. Glaude Jr
AFRICAN HISTORY John Parker and
  Richard Rathbone
AFRICAN RELIGIONS Jacob K. Olupona
AGEING Nancy A. Pachana
AGNOSTICISM Robin Le Poidevin
AGRICULTURE Paul Brassley and
  Richard Soffe
ALEXANDER THE GREAT
  Hugh Bowden
ALGEBRA Peter M. Higgins
AMERICAN HISTORY Paul S. Boyer
AMERICAN IMMIGRATION
  David A. Gerber
AMERICAN LEGAL HISTORY
  G. Edward White
AMERICAN POLITICAL HISTORY
  Donald Critchlow
AMERICAN POLITICAL PARTIES
  AND ELECTIONS L. Sandy Maisel
AMERICAN POLITICS
  Richard M. Valelly
THE AMERICAN PRESIDENCY
  Charles O. Jones
THE AMERICAN REVOLUTION
  Robert J. Allison
AMERICAN SLAVERY
  Heather Andrea Williams
THE AMERICAN WEST Stephen Aron
AMERICAN WOMEN'S HISTORY
  Susan Ware

ANAESTHESIA Aidan O'Donnell
ANARCHISM Colin Ward
ANCIENT ASSYRIA Karen Radner
ANCIENT EGYPT Ian Shaw
ANCIENT EGYPTIAN ART AND
  ARCHITECTURE Christina Riggs
ANCIENT GREECE Paul Cartledge
THE ANCIENT NEAR EAST
  Amanda H. Podany
ANCIENT PHILOSOPHY Julia Annas
ANCIENT WARFARE Harry Sidebottom
ANGELS David Albert Jones
ANGLICANISM Mark Chapman
THE ANGLO-SAXON AGE John Blair
ANIMAL BEHAVIOUR
  Tristram D. Wyatt
THE ANIMAL KINGDOM
  Peter Holland
ANIMAL RIGHTS David DeGrazia
THE ANTARCTIC Klaus Dodds
ANTISEMITISM Steven Beller
ANXIETY Daniel Freeman and
  Jason Freeman
THE APOCRYPHAL GOSPELS
  Paul Foster
ARCHAEOLOGY Paul Bahn
ARCHITECTURE Andrew Ballantyne
ARISTOCRACY William Doyle
ARISTOTLE Jonathan Barnes
ART HISTORY Dana Arnold
ART THEORY Cynthia Freeland
ASIAN AMERICAN HISTORY
  Madeline Y. Hsu
ASTROBIOLOGY David C. Catling
ASTROPHYSICS James Binney

Available soon:

For more information visit our website

www.oup.com/vsi/

Jim Bennett

NAVIGATION

A Very Short Introduction

OXFORD
UNIVERSITY PRESS

Great Clarendon Street, Oxford, OX2 6DP,
United Kingdom

Oxford University Press is a department of the University of Oxford.
It furthers the University's objective of excellence in research, scholarship,
and education by publishing worldwide. Oxford is a registered trade mark of
Oxford University Press in the UK and in certain other countries

Published in the United States of America by Oxford University Press
198 Madison Avenue, New York, NY 10016, United States of America

British Library Cataloguing in Publication Data
Data available

Library of Congress Control Number: 2016952531

ISBN 978-0-19-873371-3

Printed and bound by
CPI Group (UK) Ltd, Croydon, CR0 4YY

To Siobhán and Yolaine

Contents

List of illustrations

Chapter 1
Early navigational cultures

Navigation

The word *navigation* has at least two common senses in relation
to ships and the sea. They are different but not independent, for
they intersect and overlap. The broader of the two refers to the
entire practice of travelling in a vessel on water. This was the sense
invoked by Richard Hakluyt, for example, when he compiled and
published his *Principall Navigations, Voiages and Discoveries of
the English Nation* in 1589, and by Thomas Jefferson, when he
listed the four pillars of American prosperity as 'agriculture,
manufactures, commerce and navigation'. This book will deal
instead with the more specific and technical business of finding
the position of a ship and reckoning its progress towards an
intended goal. Early printed books on the subject might begin
with a definition similar to that used by the famous mathematician
John Dee in 1570: 'The Arte of Nauigation, demonstrateth how, by
the shortest good way, by the aptest Direction, & in the shortest
time, a sufficient Ship, betwene any two places...assigned: may be
conducted.' In its modern use, this technical sense might extend to
flying balloons or walking the moors, but here we shall deal only
with navigation at sea.

We should acknowledge also that, while this will be a story of
human navigation, other species might be judged to have superior

skills, on land, at sea, or in the air. For good reason the Institute of Navigation adopted an emblem in 1951 incorporating the Arctic tern; in the year of writing a return migratory journey by this remarkable bird from Britain to Antarctica was recorded at 96,000 km.

Navigation has helped shape the global history of humankind, but the work of the navigator in the sense defined has been, until we reach the final chapter of this book, the province of men. It has seemed awkward and even misleading to avoid the use of the male pronoun and other gender-specific words, as would be appropriate in many publications. There are exceptions, of course. A favourite of mine is the response of the celebrated geographer Sir Clements Markham to a request from the distinguished colonial administrator Sir Roger Goldsworthy for advice on learning nautical astronomy, c.1870: 'I could only refer him to an old lady in the Minories [in London], who was an excellent teacher, but who then stood almost alone.' This was certainly Mrs Janet Taylor, who gave lessons and wrote textbooks on navigation.

The story of navigation is usually begun in the Mediterranean. Yet if chronology is to drive the narrative, in truth, we do not know where to begin and might as easily choose the Indian Ocean, the South Pacific, or other regions of which we know even less. In fact the choice matters little for a historical account of navigation as a technical practice, one that traces lines of development, looking out for influences and exchange of ideas and techniques. Distinct geographies, where areas of sea were demarcated not only by land but also by climate and current, helped to create different cultures of navigation that for centuries followed individual trajectories. Only with the coincidence of a great expansion of trade that transgressed established boundaries, and the codification of navigational practice in books, charts, and instruments, can we see these technical geographies breaking down and can begin to write in terms of a congruence of navigational practice, as isolation was gradually replaced by emulation and competition.

Of course, pursuing this line of thought further back, we must reflect that these separate geographies of navigational practice were themselves theatres of movement and trade, so each culture was itself no doubt built through a spreading of knowledge and skill. Of these very early stages in the story of navigation we have little evidence and plenty of speculation. Before codification, exchange, and mutual commentary, our sources are fragmentary and much of our knowledge precarious. One thing that was shared by ancient navigational cultures, however, and that we share with them, is the sky. The sky looks different from different parts of the world—it is this feature that gives it navigational purchase—but the characteristics of its apparent movement and the principles of how its appearance changes with the observer's position are universal.

The shared sky

The sky was much more evident and familiar to the Earth's inhabitants in the great majority of time past. This is true especially of the night sky, but even in daytime, we are much less aware of all the characteristics of the Sun's cycle of motions. Most of us are more urban than our ancestors and live in a less transparent atmosphere, with our vision adjusted to an unprecedented intensity of artificial light. Time and date are supplied in ways that leave us scarcely aware that clocks and calendars are registers of astronomical phenomena. We rarely find ourselves in real darkness under an open sky. If we do, we quickly become aware of gradual changes in how the heavens appear, of their *motion*. We will speak as navigators have always done, as though the motions we observe are those of the heavenly bodies and not the Earth, as though Copernicus had not taught the world that the opposite is true.

The sky offers the regular observer registers of both direction and position, two fundamental variables of navigation (see Figure 1). We can begin with the simple observation that the Sun pursues a

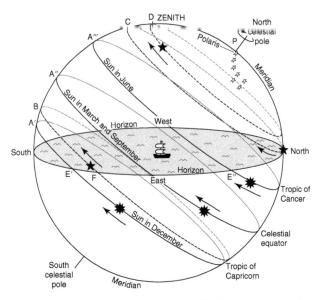

1. This diagram introduces the experience of being at sea, under the moving sky, where the sea extends to the visible horizon, above which the Sun and stars appear and below which they disappear, although no one involved in this experience thought that the Earth itself was flat. A vessel is represented at an arbitrary northern latitude, the outer circle being the intersection of its meridian (the local north–south plane) with the celestial sphere. The following clues to direction and position are mentioned in the order of their introduction in 'The shared sky'.

All bodies reach their daily maximum altitudes when crossing the observer's meridian: A′, A″, A‴ represent the Sun's culmination at different times of the year, B, C, D the culminations of representative stars.

At present Polaris in the constellation Ursa Minor is very close to the North Celestial Pole, P, and so constitutes a stationary 'North Star'.

In the latitude represented, the star culminating at C just grazes the horizon at 'North', thus giving indications of both direction and latitude.

In the latitude represented, the star culminating at B will always rise at the point F on the horizon, so indicates that direction (provided it rises at night and so is visible).

4

daily arc from east to west. It follows a symmetrical path, ascending from the horizon in the morning and falling in the afternoon. Altitude is the angle made with the horizon and it reaches a maximum when this symmetrical path is bisected, so midway between the points of rising from the east and setting towards the west precisely indicates north or south at the Sun's highest altitude. Not only the Sun, all the heavenly bodies take part in this westward motion, and all *culminate* on the *meridian* of the observer, that is, in the local north-south plane.

In fact everything in the heavens appears to be turning in circles around a point in the northern half of the sky (the North Celestial Pole) or in the southern (the South Celestial Pole); the Pole relevant to the observed path of a particular star or group (constellation) of stars will depend both on its position in the pattern of the sky and on the position of the observer on the Earth. At present the bright star Polaris in the northern constellation of Ursa Minor (the Little Bear), is well within 1° of the Pole so, in effect, marks a stationary point around which the northern

---

1. *Continued*

The Sun will rise at different points on the horizon (East, E′, E″) at different times of the year, but always generally in the east (and exactly in the east everywhere on earth at the equinoxes); the equivalent observation applies, of course, to its setting in the west and the variation in the direction of sunrise depends on latitude (there is none at the equator).

From more northerly positions a greater number of stars and constellations never set below the horizon; in the latitude represented, the star culminating at D, for example, is 'circumpolar'. If this is a star known to the sailor, the fact that it passes close to the zenith is also an indication of his latitude.

From more northerly positions, Polaris is higher in the sky; the angle between the pole, P, and the horizon, 'North', is equal to the observer's latitude.

The height of the culminating point of the Sun (A′, A″, A‴) is also an index of latitude but allowance must be made for its variation throughout the year.

celestial hemisphere appears to rotate. This was not always so. A very slow motion of the Earth's axis of rotation, completing a conical path every 26,000 years, means that the celestial Pole follows a circle in the sky at a rate of 1° in seventy-two years, bringing it at times closer to and at others farther from certain stars. When Homer wrote the *Odyssey*, for example, probably in the 8th century BC, with its references to the celestial navigation of Odysseus, Polaris was over 12° from the Pole. Other stars have at times been sufficiently close to serve as a *north star*, and even without such a clear guide, the Pole can be roughly inferred by the circles followed by nearby constellations. The need to infer in this way has always been so of the South Pole, where there is no marker star, but the circling of the Southern Cross is a valuable guide to the position of the Pole and so to the southerly direction.

Of course all this depends on having a reasonably clear sky and so is at the mercy of the weather, but from what we have seen already, we can realize that there are other indications of direction if the Pole itself is not visible. Not only the Sun but every body culminates on the meridian and, if a star is sufficiently close to the Pole not to reach the horizon in its daily path, its least altitude (lower culmination) will also mark the meridian. If a star just grazes the horizon, that too will indicate north or south. We can be more ambitious in direction finding than this, for the risings and settings of constellations will be associated with particular points on the entire horizon, which may therefore be divided in the manner of a compass, whose points can now be associated with particular stars. In that way many directions can be read in the sky and not only north and south. From a given position on Earth, a particular star will always rise at the same point on the horizon, but not at the same time, because time is determined by the position of the Sun, which moves through the stars as the year progresses. This means that for a period of the year a star will not appear at night and others must be used for the relevant direction. Further, as a star moves away from the horizon, it may take an oblique path that alters its position along the horizon (its azimuth), so that a seaman cannot follow it for long but must substitute another that has risen more recently from that horizon point. A star compass, compared with a magnetic compass, is a complex, changeable instrument; there is a great deal to learn and much to practice.

In introducing the daily path of the Sun, it was important to refer to the east and west in somewhat vague terms, because its precise intersections with the horizon depend, of course, on the time of year, as the Sun follows its annual course between the two hemispheres. It depends also on the position of the observer, but to the canny navigator, watching it day by day, it is still a directional parameter. The star compass is also affected by the observer's position. Different constellations rise and set at different points according to the observers' positions north or south (their latitudes), and the further north they stand, the greater the numbers of stars that never reach the horizon at all and are always visible in a clear night sky. This complicates the use of stars or constellations for direction finding, but the navigator turns this to advantage, for we have now strayed into that other register offered by the sky: position.

Consider first the North Celestial Pole, whether closely indicated by Polaris or not. At the North Pole on Earth it will be directly overhead (the observer's *zenith*); its altitude is 90°, and so, we might note, is the observer's latitude, or distance from the Equator expressed as an angle whose apex is the Earth's centre. Moving south, the celestial Pole will be progressively lower in the observer's sky until at the Equator it coincides with the horizon, and its altitude is zero, as is the observer's latitude. (The stars are so very far away that we can ignore completely the finite size of the Earth and treat the observer's line of sight to the celestial Pole as being parallel to the Earth's axis.) At all points between the North Pole and the Equator, the decreasing altitude of the celestial Pole is equal to the observer's latitude. Now, if we are living at an epoch when Polaris is close to the Pole, an estimate of its altitude is an estimate of our latitude. Better still, a *measure* of its altitude is a measure of our latitude. If Polaris is so far from the Pole that we should make an allowance for this, we can learn to adjust our measurement from the orientation of nearby constellations, which indicate whether Polaris is level with, above, or below the Pole in its daily circle around it.

We might pause at that last point, to note that the move to accommodate the variability of the altitude of Polaris by taking a significant amount of trouble in devising a method of correcting for this, and learning a routine for applying this correction at sea (of which more later) is a characteristic we shall encounter time and again as we follow the story of navigation. It was an impulse that would often lead navigators into ever more complicated, challenging, and time-consuming procedures and oblige them to serve long periods of training. Put simply, there was no choice, there was too much at stake simply to shrug and say, 'Well, that's too complicated'. There were overwhelming imperatives to find solutions and accept the costs in time and trouble.

As with techniques for finding direction from the sky, we can expand the repertoire for position also. While this was a convenient opportunity for me to introduce the meaning of 'latitude', that concept was not required for a viable technique for position. The position of Polaris in the sky can be associated with the position of a landfall target without invoking the angular notion of latitude. Further, the height of Polaris was only one possible observable. The position of any constellation in the sky depends on the northerly or southerly situation of the observer, and helpful indices include which star or constellation is grazing the horizon or passing directly overhead. We have seen already that the risings and settings of constellations are dependent on location, and that this is true also of the Sun. The altitude of the Sun is another index, most usefully noted at its culmination, when it is midday and the Sun is on the meridian, but that is complicated by the Sun's own movement—unlike the stars—in an annual cycle through the celestial sphere and an adjustment has to be made for the time of year. If the observed position of Polaris (when close to the Pole) changes only with the observer's location, and the positions of other stars change also with time (a daily cycle), the position of the Sun changes with location, time, and date. Early navigators adjusted their inferences and later ones would codify these adjustments into a complex but manageable package of knowledge and skill. A further clue to position north or south, which derives from the course of the Sun, is the greater variation in the lengths of day and night as we go further from the Equator.

To complete our register of the sky, what about the Moon and the planets—the other bodies with the Sun that wander through the celestial sphere? While everything culminates on the meridian, it would be only in relatively recent times that the planets were given a role in navigation, although Venus, being always close to the Sun, might indicate where it would soon rise or had recently set. The Moon, of interest first in respect of the tides, would come into its own for position-finding in the much more mathematized navigational world of the 18th and 19th centuries—something we can thankfully postpone for the present.

The Mediterranean

Not all navigation attends to the sky. If a vessel remains within sight of a familiar coastline, its progression can be noted as it passes recognized landmarks. The only navigational aid needed would be a sounding line to measure depth—the lead and line—and perhaps a written account of the course to follow and of its distinctive features, or having such information committed to memory or absorbed through experience. Later there might be a chart for the purpose. This technique is known as *coastal navigation* or *pilotage*; the term *hugging the shore* may be casually used, but this is misleading, as being close-in exposes a ship to the dangers of tides, currents, shallows, and reefs.

Knowledge of such hazards and their local occurrence is part of the skill of pilotage.

The lead and line were referred to by the Greek historian Herodotus in the 5th century BC as something his reader would readily understand. He mentions also without explanation the measurement the line gives in fathoms, while we learn that the sample of the sea bed the lead can retrieve is another clue for the experienced seaman. A fathom was the distance between the tips of outstretched arms, and later lines at least were knotted at such intervals or had a conventional series of attached markers. The lead itself came to have a standard form of a heavy but slender cone with an indentation in the base where a piece of tallow would facilitate catching a sample of the seabed.

The earliest sailing directions to survive as a text, known as the *Periplus of Scylax* (or *Pseudo-Scylax* to indicate that it was not written by a celebrated earlier navigator in the Indian Ocean after whom it was named) date from the 4th century BC. This is a record of the successive harbours, rivers, headlands, and other features, with the sailing times between them, beginning at the mouth of the Nile and proceeding west, then around the Mediterranean, and back to its point of origin. It is a document on the conduct of pilotage.

Pilotage skills would also be needed when making landfall after a voyage on the open sea. It is not surprising that some of the earliest evidence we have of such a combination of techniques comes from an island-based people in the enclosed waters of the Mediterranean—making geography again a determinant of navigation practice. The island is Crete and the people the Minoans—a Bronze Age civilization that thrived from about 2000 to 1450 BC. Seaborne trading, mostly eastwards to Egypt and Mesopotamia but also west as far as Spain, took the Minoans out of sight of land for several nights at a time, when the sky must have been their principal guide; however, if they identified an individual pole star, it would have been Draco at their epoch, not Polaris.

With the contingencies of geography still in mind, we might mention also that the annual cycle of seasonal change in the Mediterranean meant that, according to a commonly cited rule, only the summer months brought the favourable seas and reliable winds that were judged to make sailing worth the risk. However, this applied more strictly to coastal than to open-sea routes, while sailing plans were also obliged to accommodate seasonal change in the prevailing winds. A north wind, for example, would facilitate a voyage from Phoenicia to Egypt but might be hoped for only between late September and December.

The Mycenaeans of southern Greece were also successful traders in the eastern Mediterranean, benefiting from the decline of the Minoans in the mid-second millennium. They too disappeared in the general waning of Bronze Age cultures in the region and the great sea-going merchants who rose in their place, the Phoenicians, based around Tyre and Sidon on the eastern edge of the Mediterranean, outshone both with their adventurous mastery of the sea.

Encouraged by Egyptian allies, for example, a Phoenician expedition beginning in the Red Sea accomplished a three-year circumnavigation of Africa in around 600 BC. Other ancient feats of navigation began in Carthage on the north coast of Africa, originally a Phoenician port but gaining independence in the mid-7th century. The Carthaginians had trading settlements as far as the Atlantic coast of Africa, and on a celebrated expedition to service and extend these, Hanno the Navigator sailed on as far as the Gulf of Guinea, though quite where he turned for home is debated on the basis of a surviving transcription of his account of the voyage. This was in the 5th century, when also the Carthaginian explorer Himilco, recorded in a later Roman source, sailed north on leaving the Mediterranean, reaching Portugal, France, and England.

In the following century, the Greek explorer Euthymenes of Massalia (Marseilles) sailed south to Senegal, and perhaps as far as Ghana, while from the same city, Pytheas began a famous journey north. He wrote an account that became widely known, so that we have secondary references, even though a copy of the text itself has not survived. These references tell of Pytheas reaching what we would call the Arctic Circle and witnessing the midnight Sun. His achievement is now understood not as a single voyage but as a series of seaborne and overland stages, using local vessels and seamen as appropriate, which speaks to the skill of sailors from the islands north of Britain—of Orkney and Shetland. It must have been with their help that Pytheas reached the Island of Thule, whose most likely modern identity is Iceland.

These extraordinary voyages are well known but what can we say about the navigational techniques they employed? Even though it would be unknown on the outward voyage, tracking the coast is not irrelevant, as it establishes a path that can be retraced, but other clues to position would be needed. One cannot even say that a venture of circumnavigation could be assured, barring disaster, to return to its starting point, as it was not believed that Africa had clear sea to the south. Herodotus did not believe the story of the Phoenicians' voyage for this reason, but neither did he accept the detail of their story that both makes it credible and points to the sky as one of their guides. When rounding the southern cape the sailors found that the Sun was on their right-hand side.

That these early navigators used the sky for guidance can scarcely be doubted. Homer has Odysseus using the stars for direction by keeping the Pleiades and Ursa Major on his left side while sailing east. Callimachus, a poet and scholar at the Library of Alexandria, says that the Phoenicians sailed by Ursa Minor, while his contemporary poet Aratus gives the explanation that it was closer to the Pole than Ursa Major. At the end of the pre-Christian era, the geographer Strabo also accepted the importance of Ursa Minor for the Phoenician sailors.

An ability to recognize the characteristics of a set of winds in the Mediterranean, blowing from particular directions, was the basis of a different division of the horizon—less accurate than the stars but a useful secondary guide. Sailors for the most part settled on an eight-fold division, which curiously survived longer as a convention on Italian surveying instruments than in navigation.

As well as direction, there is evidence of the sky being used for position. Pytheas used the elevation of the Pole as a measure of his progress northwards, pointing out that there was no star to mark the Pole (and taking issue with the astronomer Eudoxus on this point), which had to be inferred from the positions of three circumpolar stars. Pytheas also noted the changing population of circumpolar stars and variations in the length of daylight—two further indications of changing latitude.

We have not yet been able to cite an instrument being used for either direction or position. While it would seem an obvious expedient, the first recorded mention of sailors measuring the heights of stars by the masts and spars of their ship comes from the Roman poet Lucan in around AD 65, but he does attribute the practice to the Phoenicians.

The Pacific

A history of the techniques of navigation must interweave chronology and culture, but for a manageable narrative the weave has to be loose. Given coexisting navigational cultures and their distinctive geographies, it would be perplexing to follow a strict chronology, yet the narrative is marked by moments of encounter, which may or may not lead to significant influence. If early navigation in the Mediterranean, its large space almost entirely enclosed, can be constructed as a 'likely story' emerging from prehistoric times, the same is true of the very different geography of Oceania. Here, by contrast, is a vast expanse of sea, with many scattered islands and little in the way of natural boundary. Yet even the geography is construed culturally as well as physically: where Europeans might see these far-flung islands separated by the ocean, Pacific peoples might insist that this is how they are connected.

Of the three broad areas recognized as Oceania, Melanesia is closest to Australia and the first group of islands to be inhabited, extending from Papua New Guinea to Fiji and including the Solomon Islands and New Caledonia. To the north is Micronesia, which includes the Caroline and Marshall Islands. Extending far into the east, Polynesia reaches to Easter Island, as well as north to Hawaii and south to New Zealand: a vast area that can be outlined as a triangle whose sides are some 4,000 miles in length. It contains over 1,000 islands where similarities of language and culture indicate a long period of associated migration and continuing contact. It is thought that Tonga and Samoa were inhabited about 1000 BC and Easter Island around AD 1200. Of these regions, it is the Micronesians and Polynesians who were responsible for the most outstanding feats of navigation—so outstanding that Europeans have sometimes found it difficult to accept that they were achieved on purpose and not by chance.

How and when this extraordinary migration of people could have happened is a matter for continuing discussion but here we need consider only what can be said of the navigational techniques that were used. Much of what we know about this comes from the survival until recent times of indigenous navigation and the dedication of adventurous scholar-seamen who have made it their business to learn and record what is still known and practised, even if this is almost certainly a reduced and incomplete picture of this navigational culture at its height.

Navigation in Oceania was practised without instruments and with no written accounts. Whether the so-called *stick charts* of the Marshall Islanders constitute instruments for use at sea is debatable: skeletal structures of twigs and shells, indicating directions between islands and patterns of sea swell, they are more instructional or mnemonic aids than used for following a course in the manner of a sea chart. Perhaps the only other candidates for instruments would be pennants for indicating wind direction. Without a written corpus, navigators constituted a social elite, trained from an early age, and committing much to memory through both instruction and experience. The archive of knowledge was vested in people rather than paper, and the navigators guarded it closely.

There was much to learn. First the sky had to be mastered—the stars and constellations of the sky compass; their sequences through the night; and the alternatives for different seasons. This must be combined with bearings from one island to another, and the associated stars, together with whatever routes or sequences of bearings that were manageable for longer voyages. Zenith stars associated with particular islands had to be learnt. Other directional information came from sea swell and prevailing winds, the daytime path of the Sun, or the flight of migratory birds. Bird species needed to be recognized and their habits known, including migration patterns and when certain land-nesting species could indicate the direction of an island over the horizon.

Not all memory is cerebral. Recognizing swells and winds was as much a matter of feeling as knowing. Even more difficult was adjusting for invisible currents, and the heading of the vessel had to be set to make allowance for this as well as for wind. Estimates of speed—essential for any determination of position by *dead reckoning* (an inference from the course steered for a given time)—were a matter of experience and, to some extent, so too were estimates of time, though here there were checks that could be made in the sky.

Making landfall called on other knowledge. Here navigators could not rely on the assurance of encountering a coast and recognizing their position along it, but even when the target was an isolated island, there were clues to nearby unseen land. We have mentioned the behaviour of birds, but swells and waves were also altered by land, while clouds could differ in colour and formation according to whether or not land was near. Such are only a smattering of the techniques at play, while their integration and comparative assessment in making vital judgements and decisions must have relied on long experience.

Before the compass

It has been possible to say something about ancient navigational techniques in the Mediterranean and the Pacific for different reasons: in the former case because we have fragmentary literary records, in the latter because a residual practice survived long enough to be learnt and understood in recent times. In both cases our accounts can be supported by arguments from what must have been noticed in voyages on an open sea under an open sky.

Seafaring was taking place elsewhere as well. Even allowing for differences in the configurations of land masses, such as the existence of a *Bering land bridge* between Asia and North America, and lower sea levels than today, early human migrations required sea crossings. Yet it seems impossible to say if or when anything that would qualify for our purposes as 'navigation' was involved. From further south on the eastern Asian seaboard, organized seafaring around the southern shores of China was established by the second millennium BC and in about AD 350 Chinese ships reached Malaya, and then Ceylon by the end of the century.

The Indian Ocean introduces a new characteristic to our narrative. So far, navigational cultures have been largely separate, certainly in the Pacific through isolation, but also effectively in the Mediterranean, where a succession of traditions interacted, mainly through conflict and displacement. The Romans became a great maritime power, for example, over the course of a lengthy succession of wars with the Carthaginians. There is little to say in this context about developments in the techniques of navigation, though much has been written about the changing design of warships. In the Indian Ocean, however, navigational cultures overlapped, creating the possibility and occasionally the reality of exchange. This might be taken as the third great theatre for navigational exploits by maritime peoples that led to sustained seafaring cultures. Here there were different societies in action—Egyptians, Romans, Chinese, Indians, and Arabs—with corresponding occasions for technical transfer.

From at least the third millennium BC the Egyptians, as well as sailing in the Mediterranean, were organizing expeditions south in the Red Sea to the 'Land of Punt', probably in the Horn of Africa, trading for frankincense, ebony, and myrrh. What might be said of Egyptian navigational technique? Illustrations of ships as well as texts indicate three navigational specialists in a ship's crew: one was *the man at the prow*, who had a sounding rod (more constantly in use on river vessels) and kept an eye on wind, waves, landmarks, and sky; and another, *the man at the poop*, who had charge of the rudder; both of them under the command of the navigator proper. As well as directing the other two, we learn from a tomb inscription that the third specialist, the navigator, paid constant attention to the direction of the wind and the management of the sails and rigging.

After the death of Alexander the Great in 323 BC, the Ptolemies not only became confident in navigating the Red Sea but established a maritime trade with India. According to Posidonius, however, in an account repeated by Strabo (which Strabo himself did not credit), this began with the rescue in the Red Sea of a ship's pilot from India, and the mounting of an expedition in 118 BC, led by the Greek navigator Eudoxus of Cyzicus, sponsored by the Egyptian King Ptolemy VIII and guided by this unnamed Indian seaman.

So, encounter was key to this navigational development, and 'navigational' it was, as it required knowledge and application of the monsoon winds. This annual cycle of south-westerlies in summer and north-easterlies in winter is another example of navigation shaped by geography. It set a pattern for sailing between India and East Africa or South-East Asia that lasted till the age of steam. Whether or not the practice was revealed through a rescued seaman, it was to be actively promoted by the Greeks and Egyptians, and later by Roman ships. And there is no reason to reject the story's underlying assumption that there was already an established practice among Indian sailors, for there was Indian maritime trade along the coast of Arabia, including the exchange of goods with Greek ships in Eudaemon (Aden) in the 2nd century BC.

We should mention a technique, reported by Pliny, as practised by seamen from Ceylon, who would release shore-sighting birds so as to note the direction they choose to fly, or whether they opted to return to the ship. In fact this is reported widely in early seafaring on the open ocean, and could equally be mentioned in respect of the Pacific or the Atlantic.

Arab and Chinese seamen were the principal navigational players in the Indian Ocean in the medieval period, when we find perhaps the most famous example of the transmission of a navigational technique. The appearance of the magnetic compass at sea was surely a transfer of practice from the Chinese to, in the first instance, the Arabs. Though we have neither dates nor names to give it a precise occasion, such was the far-reaching importance of this encounter that we shall take it as the start of a new era requiring a new chapter, where we shall also deal with Chinese and Arab navigation more generally. Before we do so, however, there is one further theatre that must be mentioned, where the early navigational cultures remained untouched by the magnetic compass and where, with the universal exception of the sky, conditions for seafaring were profoundly different, namely the North Atlantic.

The Irish of the early Middle Ages introduce us to a motive for deep-sea navigation we have not yet mentioned, namely the devotional, whether adopted from a missionary or an ascetic impulse. Their *immrama* are fabulous tales from the 8th century of adventurous voyages of discovery and self-discovery, of which the most famous is that of St Brendan the Navigator, who lived in the 6th century. The earliest account of his voyage dates from *c.*900. Whatever basis such tales have in historical fact, they certainly reflect a respect for seaborne exploits, and they are surely related to an esteem for distant monastic settlement or individual hermitic ambition. Examples grounded in more substantial records include St Columba's voyage to Iona in 563, followed by Irish settlements in the Orkney and Shetland Islands, and on the Faroes by around 700. There are plausible grounds for accepting that Irish monks were established in Iceland by the end of the 8th century. All of this exploratory voyaging was before the era of the Vikings, who displaced the Irish from the Faroes in around 800 and probably from Iceland later in the century.

We can say very little about any navigational techniques underpinning these voyages, but in such northern latitudes especially it is impossible not to imagine that the sky was the principal guide. It is thought that the beginning of the year (by our reckoning) would have favoured navigation by the stars, avoiding the luminous skies of the summer months. The account of the Irish geographer Dicuil, writing in *c.*825, suggests that the band of Irish monks who first landed on 'Thile' (often taken to be Iceland, which we have met in the account of Pytheas as 'Thule') did so in the depth of winter. He records their observation that at the summer solstice there the Sun only just dipped below the horizon. Regular traffic, such as was established to the Faroes, will have made the most also of the kind of sensitivities we noted as being utilized in the Pacific—to the flight of birds, the sound of waves on a strand, and so on—informing above all the skill of the helmsman.

If we have allowed the Irish to introduce a religious motive for navigation, we can acknowledge another omission through the Norse, namely fishing. The early experience of Norse seamen came, as it would later for the English, through Atlantic fishing, before they became known for trading and even better known for marauding. They too had to gain knowledge of coastal landmarks, winds, tidal streams, currents, the flight of birds, the changing colour of the sea over underwater banks, and so on. Their exploits were not limited to the Atlantic, as they brought their vessels to London, Dublin, Paris, and Constantinople. But it was in northern latitudes that their navigational skill was tested to the extreme. They too crossed the sea to the Orkneys and Shetlands, the Faroes, and Iceland, and then on to Greenland and North America or 'Vinland', which they reached in the 10th century.

On the outward voyage, keeping the Pole Star to starboard and the midday Sun (occasionally the midnight Sun) to port would have given them direction. Maintaining a reckoning as best they could would have been vital, while looking out for such clues as the sea colour over the Viking-Bergen Bank, the behaviour of birds, the occurrence of whales, and so on. There are references in later medieval Norwegian texts to the use of the sounding lead in the earlier voyages.

As for position, north and south, over the course of their seafaring exploits while attending to the sky for direction, the Vikings cannot have failed to notice some of the changes we have mentioned earlier in this chapter—in the height of Polaris, for example, or the pattern of movement of circumpolar constellations, or the varying length of the day. For many of the voyages into or across the Atlantic, however, or at least for long sections of them, maintaining a northerly station while sailing east or west would have ensured success. A vessel might depart the Norwegian coast from a known point and endeavour to keep, say, Polaris at roughly the same height in the sky. No concept of latitude as such would be involved in this practice, but it is equivalent to what would later be known as *latitude sailing*. Not only does this seem a natural evolution from maintaining an orientation to Polaris for direction, it seems to be present in the saga literature that becomes available from the 13th century.

From the same sources we have sailing directions such as *keeping the sea*, that is, the horizon; *half-way up* a mountain slope when passing to the south of the Faroes, or while sailing a stretch of the Norwegian coast; and many references to birds, whales, narwhals, and coastal landmarks. One object mentioned, however, has attracted particular attention in recent years, the *sunstone*, which some have interpreted as being a crystal of the mineral Iceland spar, whose light-polarizing properties could have been used to detect the position of the Sun in cloudy weather, and so to find direction at sea. If this was so, we have an instance of something that has been absent so far from the story of navigation—the use of an instrument—but that will be prominent in what is to follow.

Chapter 2
Medieval and Renaissance learning and practice

The Indian Ocean and the Arabs

Using the broadest brush, the maritime history of the Indian Ocean in the medieval period can be told through the successive fortunes of the mercantile activities of Chinese and Arab seamen. I have mentioned Chinese trade with Ceylon by the end of the 4th century, and soon afterwards they were sailing in the Persian Gulf. A decline in the Chinese presence in the Indian Ocean from the 9th century was matched by a rise in number of Islamic Arab seamen, and in the same period there were significant Arab *factories* established for mercantile trade in Guangdong. The Chinese began to re-establish their importance in the Pacific from the 13th century, and under the Ming dynasty in the 15th they were sailing to the west coast of India and to East Africa. It was in the same epoch that Europeans first sailed into the Indian Ocean from the west, though this was towards the end of the century, after the Chinese had disengaged.

Given the presence of three major cultures of ambitious development in navigation—Chinese, Arab, and European—it is not surprising that historians contend, for the most part politely, over innovations, priorities, and sources. For our part, we can describe the techniques characteristic of the different players, without setting too much store by priority as such and while noting that this was generally (apart from the eventual decline in Chinese activity) a time of convergence.

We can begin with techniques used by Arab sailors. We have seen that the division of the horizon by points or zones marked by the rising or setting of stars or constellations was practised generally at sea. In the Indian Ocean it was part of the learned experience of the Indians, Chinese, and Arabs, and it was here that it arrived at a highly developed system of thirty-two points that became known as the *Arab compass*. Horizon astronomy was used for setting and maintaining bearings, with all the complications of hourly, seasonal, and locational variation that we noted for sailors in the Pacific. There was one sky.

Identifying a star on the horizon could be a clue to position as well as direction. For a given orientation of the revolving sky the stars seen close to the horizon would depend on the position of the observer on Earth. In the West it was customary to divide the circle of the ecliptic (the annual path of the Sun through the celestial sphere) into the twelve zodiacal signs still familiar today, each occupied by a constellation—Aries, Taurus, and so on. Arab astronomers and astrologers (the distinction is somewhat artificial at this time) had an alternative division into twenty-eight *lunar mansions*, again identified by their constellations. This astronomy could be applied at sea for finding latitude: the identity of the constellation (lunar mansion) on the observer's meridian, found by recognizing which of them was culminating, was associated with different known stars on the horizon, according to the observer's latitude.

Moving away from the horizon itself, we have already mentioned the practice of relating position north or south to the height of heavenly bodies above the horizon, in the simplest case to the elevation of the Pole Star. It seems likely that the Arabs were the first to devise an instrument specifically for registering this change at sea. This was the *kamāl*.

Astronomers had long been familiar with the relationship between altitude measurements and latitude. Some of their instruments that had to be directed to the Pole, such as the armillary sphere or the equinoctial sundial, were designed with adjustments to make allowance for latitude. Others, notably the cross-staff, were for measuring angles in the sky, including and probably most conveniently, altitudes. In the case of the astrolabe, an altitude measurement was the main function of the back of the instrument, for it comprised a circular degree scale, where zero represented the horizontal plane, and a suspension ring at 90° for hanging it vertically, as a centrally pivoted sighting rule or *alidade* was trained on the target. That operation would immediately give the altitude of the target, a measurement to be used in working the functions of the front side of the astrolabe.

While Islamic Arabs established a long and distinguished tradition in the use and development of the astrolabe, beginning in the 8th century, they did not adapt it for use in navigation. Neither was the cross-staff, the most directly applicable altitude instrument of the astronomers, the source for the kamāl. Instead we might begin with the simple expedient of gauging the height of Polaris, its elevation above the horizon, by stretching out an arm and seeing how many fingers' breadths covered the space between the horizon and the star. Certainly that was a familiar method used in the Indian Ocean by sailors from different lands. One way of making the practice more general and reliable would be to hold out wooden tablets of a range of sizes, and the use of such tablets at sea is recorded from the later 9th century. A string might be used, held in the teeth and attached to the centre of the tablet, to steady it as a judgment is made. A final step to the kamāl shortly afterwards was to knot such a string along its length and, while keeping it taut, see at how many knots' length of string the tablet just spanned the distance between the horizon and the Pole Star.

As we noted with the more primitive practice suggested for the Vikings, no concept of latitude needed to be involved. A known number of fingers, one of a set of tablets, or one of a series of knots could indicate the point where a change of course should be made for a particular route or could be a register to be maintained for a planned landfall. But this was a profound step, the definite introduction of an instrument, something that would come to characterize the practice of navigation.

Neither were we far from adding another and, as it would turn out, more iconic instrument to the routine of the navigator, namely the magnetic compass. But first a few further points about Arab navigation. It should be stated that obviously Polaris would be of limited use when sailing south of the Equator, although with experience its position might at times be inferred from the other stars in Ursa Minor. Otherwise, the culminations of other selected stars must be observed. Then there is the problem of the displacement of Polaris from the Pole—almost 4°, for example, in *c.*1400. Arab sailors came to learn an adjustment that, here again, depended on knowing the lunar mansions and recognizing their constellations. The seaman would observe which constellation was culminating (i.e. was on his meridian) and, since this was an index of the orientation of the sky and therefore the position of Polaris with respect to the Pole, it could be associated with an adjustment to apply to the measured altitude of Polaris to give the height of the Pole. (In fact it could be customary to treat the minimum altitude of the Pole Star as the reference point to which the observed altitude was adjusted, rather than the Pole itself). Here also we are not necessarily thinking of measurement in the degrees of the geometers. If Polaris was not visible, other stars could be used, requiring more complex routines.

The famous navigator of the Indian Ocean, Sulaymān al-Mahrī, from the early 16th century, describes the practice of latitude sailing, but it was surely at play significantly earlier, being a natural consequence of finding position north or south by the elevation of Polaris. Sulaymān describes deep-sea voyages where the course is altered out of sight of land, guided by the height of the Pole Star or, if necessary, some known constellation.

Medieval Arab sailors were active in the Mediterranean as well, but the Indian Ocean posed the greater challenge for navigation, with a much larger, more open area of sea, and a greater range of latitudes. Not that the techniques of pilotage were redundant there, but the temptation to find more efficient alternatives to long coastal voyages and the opportunities offered by the predictable pattern of seasonal Monsoon winds drew navigators, proud of the greater skills required for such ventures, into and across the ocean. They sailed to the east and west coasts of India, to Burma, Sumatra, and China, and to the east coast of Africa.

The Chinese and the early compass

The Chinese use of the directional properties of the magnet in land-based geomancy made the technique publicly known from the 1st century AD, and in the nature of the subject there may have been an earlier secret knowledge. It was discovered that magnetic properties could be induced in iron needles by stroking them with a *lodestone* (or *loadstone*), a naturally occurring magnet formed of the mineral *magnetite*. Floating such a needle in a bowl of water provided the first form of mariner's compass and the earliest Chinese description of this practice dates from just before 1044.

From there it is not at all clear or certain how and when the compass may have spread to other navigational cultures—whether, for example, the route was that of overland trade or through encounter in the Indian Ocean, or even whether independent discovery played a part. This is a perilous topic for an introductory book. There are claims, for example, of Tamil records from the 4th century of a floating fish-shaped magnet, with linguistic indications of a Chinese source. It would be wrong to give an impression of assurance here and we might best fall back on the authority of the celebrated Sinologist Joseph Needham for a solidly attested use on Chinese ships in 1090 and the likely use by mariners from the 9th or 10th century.

The first European record is an account by the English theologian Alexander Neckam, writing at the end of the 12th century. Working in either Oxford or Chichester, this scholarly monk had no obvious link to navigational novelty and he mentions the practice as though it was fairly standard. He had spent some years studying and teaching in Paris around 1170, so may have heard of this in a context closer to the Mediterranean diaspora or could have witnessed its use when crossing the English Channel. The earliest extant Arabic descriptions date from the 13th century and it is then that we hear of Islamic navigators in the Indian Ocean using a compass, including one with a floating needle shaped like a fish. However references survive to earlier tracts which are no longer extant.

What, then, can be said of the introduction of the compass to the practice of navigation as a vital source for direction in cloudy weather? Chinese sailors were first to have this advantage; it was known to European seamen in the Mediterranean and probably outside in the 12th century, and it is recorded in the Indian Ocean from the 13th and may have been in use there by seamen other than the Chinese from considerably earlier.

The wet compass with a floating needle, using the lightest of floats such as a straw, was not the only form and in the 12th century the Chinese had instruments where the needle was balanced on a pivot pin. Before moving to developments more characteristic of the Mediterranean, we must acknowledge the most ambitious and remarkable of Chinese navigational ventures, no doubt enabled in part by the advance in technique embodied in the magnetic compass.

A change in Chinese attitudes to distant voyages for trade came in the Song dynasty, and specifically under the 12th-century

Emperor Gaozong, when fresh studies of navigation, drawing on Arab and Indian sources, were undertaken, information on coastlines, tides, and currents was collected, charts were drawn, and sea-trade encouraged. By the early 13th century the Chinese were again a maritime power in the Pacific and Indian Oceans, in the latter case winning back control of trade from the Arabs.

It was in the first half of the 15th century that this Chinese revival reached its acme. Between 1405 and 1433 seven great treasure fleets sailed successfully for trade and for the glory of the Emperor, to Sumatra, Ceylon, India, the Persian Gulf, the Red Sea, and the coast of East Africa. The initiatives came from the Ming emperors, Zhu Di and Xuan Zong, and command for most of the voyages fell to a man of outstanding resourcefulness and skill, court eunuch Zheng He.

Navigation

These extraordinary expeditions are famous for their magnificence and opulence as well as their daring ambition, but how much do we know about the navigation involved? The magnetic compass was now a vital tool, while for the improvement of dead reckoning, measurements were taken using what came to be known in the West as the *Dutchman's log*. This provided a measure of the ship's speed, in the Chinese application by throwing a floating object, the *log*, into the sea at the bow, and walking in pace with it as the ship passed, chanting a rhyme to measure the time interval. This could be converted into a measure of speed to be applied to the duration of the watch, itself measured by the burning of incense sticks. The Chinese also used that ancient tool of pilotage, the sounding lead, prepared for sampling the bottom.

Dead reckoning depends either on prior experience or some form of sailing directions or chart. Zheng He had a long scroll, unrolled to the section corresponding to the fleet's position. This was not a scaled map but a representation of the unfolding coastlines, with compass bearings, distances, coastal features, rocks, shoals, and soundings. And something else—alongside significant ports were the altitudes of the Pole Star in *finger-breadths*. These measurements were not taken directly with fingers, since the technique had been standardized into a set of twelve wooden tablets held at arm's length. This was, of course, one of the methods used in the same ocean by Indians and Arabs, but when the Chinese employed the more sophisticated techniques of the sky compass, they were deploying their own arrangement of constellations.

With all these ingredients in place—not only the compass, the lead, and the log, but values for polar altitudes linked to target ports, and an instrument to measure and help maintain a course for a given altitude—the Chinese navigators could choose between coastal and latitude sailing, as suited their voyage. They could traverse the Indian Ocean with efficiency and confidence. Giraffes could be brought from Malindi to Beijing. Yet when the Portuguese entered the Indian Ocean in 1498, this epoch had passed. In Malindi, Vasco da Gama met Indian traders and Arab navigators but no Chinese.

Charting the Mediterranean

Chinese and Arab navigators, sailing mainly in the Indian Ocean, had made significant instrumental innovations, but one critical and long-lasting development seems to have been due to Europeans in the Mediterranean. This was the use of a sea chart as an instrument—not simply a guide for consultation, a documentary record, or a didactic tool, but as an instrument to be manipulated and used. The enclosed geography of the Mediterranean, the use of sets of sailing directions recording distances and sometimes directions (a famous example being the Italian *Lo Compasso da Navigare*, surviving in a manuscript of 1296), and the introduction of the magnetic compass all favoured the emergence of a distinctive navigational technique, underpinned by the sea chart. It is worth noting that the compass and the chart appeared in the Mediterranean at roughly the same time. This technique can be called *bearing and distance* sailing and was essentially a developed form of dead reckoning.

The medieval sea chart of the Mediterranean has long been admired as an extraordinary feat of cartography, quite apart from its status as a tool of navigation, though in fact these two aspects are closely linked. Contemporary land maps, in the *mappa mundi* tradition, are concerned with a narrative of human destiny and devotion, and look naive and hopelessly inaccurate if viewed with a modern interest in topography. The sea charts, in striking contrast, present us immediately with the familiar shorelines of the Mediterranean and the Black Sea and, depending on their extent, of the north-west coast of Africa, the Iberian Peninsula, France, the Netherlands, Denmark, and even the British Isles. The source of their evident accuracy is the experience of seamen and their use on ships for practical navigation. Not only do they have to work, but it is in the interests of everyone concerned—makers and users—that over time they should be improved.

These are *portolan charts*, their name taken (some think unfortunately) from *portolano*, Italian for a set of sailing directions, the earliest surviving chart being the *Carte Pisane* at the Bibliothèque nationale in Paris, dating from the late 13th century. Their heyday was the 14th and 15th centuries, made mainly by Genoese, Venetian, and Catalan hydrographers. They are drawn in ink—black and coloured—on vellum, which often preserves some of the shape of the original skin. Some were made for record, instruction, or decoration, and these will now be extant in disproportionately large numbers, but most were intended for practical use, and the vast majority of these will not have survived the ravages of life at sea.

One of the most evident features of a portolan chart is clusters of straight lines intersecting as they radiate out from focal points, or *wind roses*, distributed across the surface in a systematic pattern. There are generally sixteen and later thirty-two directional lines, known as *rhumb lines*, coming from each rose. They are based on the Mediterranean division of the horizon into eight wind directions, which had to be accommodated to the fundamentally astronomical orientations, originating in the sky, of north, south, east, and west. The lines for the eight winds will often be marked by the initials of their names ('T' for Tramontana, 'G' for Greco, and so on) and drawn in black or brown, lines for eight half-winds between them in green, and the sixteen quarter-winds in red, to help trace the chosen direction.

Other features are a scale bar, though the underlying unit is not specified, and huddled names of coastal features, written at right angles to the shoreline on the inland side, with the more significant places in red ink. Black dots or a cross indicate rocks; red dots, sandy shallows. Careful examination has shown that the matrix of roses and intersecting rhumb lines was set down before the shorelines were drawn and the names added.

A slightly later treatment of the originating points for the rhumb lines was the *compass rose* (see Figure 2), probably introduced by Catalan makers. The earliest surviving example dates from 1375 (preserved in the 'Catalan Atlas', another cartographic treasure of the Bibliothèque nationale). Here a star-like diagram of eight or sixteen compass points occupied at least some of the nodes for the rhumb lines. Since these lines are evidently the courses followed by maintaining a particular compass bearing, an association with the magnetic compass is a natural one. But the connection is closer still. The points of the rose came to be more commonly marked, not with wind names, but with the northern European nomenclature based on north, south, east, and west. This was becoming the standard terminology of the seaman, perhaps because compound names (north-east, north-north-east, etc.) were more readily created for the intermediate points than would have been the case with wind names. For reciting the thirty-two compass points, as mariners learnt to do, or *boxing the compass* as the English put it, four monosyllabic terms provided a good basis. To appreciate fully the

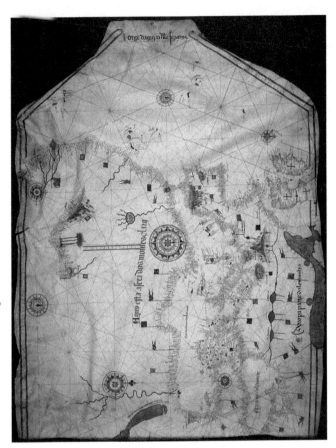

2. Portuguese portolan chart of 1492 with compass roses and radiating rhumb lines.

link between the compass rose on the chart and the compass, we must first return to the story of the magnetic compass itself.

I have mentioned the floating magnetized needle and the needle balanced on a pin, perhaps with the help of a bearing attached to the needle. A development from this was to fix the needle and bearing

32

to the underside of a circular card or *fly*, on which were marked the cardinal and intermediate directions in the form of a rose, with north aligned with the needle. The needle itself was thus concealed from view and the user saw a compass rose with not only north but all the cardinal directions immediately evident, and bisecting these the *ordinal* points, and perhaps further sub-divisions.

When did this happen? Did the compass rose appear first on the portolan chart or the mariner's compass? A transfer in either direction would be plausible but we have earlier instances on surviving charts than on compasses. On the other hand, although some 180 charts survive from between 1300 and 1500, we have no marine compasses at all before the later 16th century, even though we have records of their use at sea. Without further evidence, we cannot say when the compass rose was set above the magnetic needle and whether its distinctive but enormously varied design was first seen on a chart or a compass. The more common assumption is in favour of the chart, but that may simply reflect a greater following at present for the history of cartography. The historian of navigation and practical mathematics, E.G.R. Taylor, argued firmly for the compass. Whichever is true, the shared use of the compass rose is indicative of the close relationship in the practice of contemporary navigation between the magnetic compass and the portolan chart. They are emphatically linked by an immediate visual coherence.

Bearing and distance

How was the portolan chart used in navigation? We can first think a little further about the use of the magnetic compass. Clearly the needle floating in water had scarcely any relevance to the application of such a chart for navigation, based as it was on the assumption of a choice between thirty-two rhumbs. Setting a rose on a compass card above a pivoted needle, however, was immediately applicable. The fixed box or *binnacle* containing the compass had a prominent mark, called the *lubber-line*, orientated with the prow of the ship, so the helmsman's task was to keep this mark aligned with the point on the card representing the bearing to be sailed.

Does this help us place the historical link between the compass rose on the compass and on the chart? Given that a divided circle of some kind is necessary, the other, intermediate development from the floating needle might have been to pivot it above a fixed card—the arrangement used for land-based compasses, for example in surveying, and used for early portable sundials, which need to be orientated to the meridian. Adopting this arrangement at sea the helmsman would have first to turn his binnacle so that the north point on the card was directly underneath the needle, then find the bearing to be sailed on the card and mark this with a temporary lubber-line. He would then have to turn the binnacle and fix it somehow so that the lubber-line was aligned with the prow, and then steer so as to keep the needle over the north point on the rose. This would be a complicated procedure, involving the re-alignment of the binnacle on the ship with every change of course, which would be frequent on a sailing vessel being blown even slightly off-course or needing to tack before a contrary wind. Further, there would be a significant error of parallax involved with any oblique reading of the needle's position above the card. Given the evident ambition represented by the thirty-two-point compass rose, it seems very unlikely that setting the magnetic needle above it was a technique that had much, if any, use at sea. And given the need to match the ambitions of chart and compass in a single navigational technique, a plausible conclusion is that the wind rose on the chart and the compass rose set above the needle were roughly contemporaneous, and the compass rose was incorporated into the later charts to reflect how the technique was realized at sea.

If the reliance of this argument on the use of a thirty-two-point division of the compass seems doubtful (notwithstanding its presence on the portolan charts) in the absence of surviving compasses, it is worth remembering its appearance in Chaucer's famous *Treatise on the Astrolabe*. Written in *c*.1391 for a 10-year-old boy, there is an incidental reference to the seaman's division of thirty-two points as a commonplace aside: 'Now is thyn orisonte departed in 24 parties...al-be-it so that shipmen rikne thilke partiez in 32.'

Using the chart would require a rule and a pair of dividers (often called a pair of compasses in this context). The navigator would lay his rule on the course he wished to follow, as represented by the chart—often the line between two ports—and locate a rhumb line that was parallel to this, or as close as he could find, and, tracing this back to the originating wind rose, determine which bearing he had to follow. Opening his dividers to a convenient length on the scale bar and stepping this off along his rule would yield a rough determination of distance. While sailing on the required bearing, speed might be estimated and, from a measure of time with a sand-glass, a reckoning made of distance covered (in nautical terms, *made good*). The earliest reference to a nautical sand-glass comes from the 14th century, but they may well have been used earlier than this. And we must not forget that the lead and line was a continuing necessity, especially close to shore. There were further refinements to the *bearing and distance* approach to dead reckoning that we can date securely only to a slightly later period, so we shall return to this in time. For now, we can mention that the compass card quickly adopted some conventional iconography: a *fleur-de-lys* to indicate north (a very long survivor) and a cross for east.

One further *instrument*, in the form of a set of tables, known as the *toleta*, was available to some navigators from at least the 14th century. The chart might indicate what direction should be sailed for what distance, but wind and weather might well enforce something different. It would be necessary both to keep a record of the course that had actually been followed and decide how to proceed onwards, so as to regain the intended course. The earliest surviving set of such tables dates from 1428, but the Catalan mathematician and philosopher Raymond Lull described the use of something similar in *c.*1286. In an encyclopaedic work of instruction (though not the instruction of seamen) he explains that sailors have the use of a chart, a magnetic compass, and an 'instrument', which is the table he describes. For given numbers of quarter-winds sailed off the intended course (the use of quarter-winds implies a thirty-two-point compass) and for given distances, the tables yields the reduction in the distance from the destination and the distance from the originally intended course. A second table of the *toleta*, though not described by Lull, gave the distance that would be sailed to intersect with the intended course and the corresponding distance by which the true course would then be advanced for each of the eight quarters between the original rhumb (i.e. a parallel course) and one at right angles (i.e. sailing directly back to the intended line rather than approaching it obliquely).

The earliest surviving *toleta* is from the 15th century and offers a graphical method in the form of a mathematical instrument on paper, but with either the tables or the geometrical construction, this would have been a complicated business. For many sailors it would surely have appeared to belong to a distant, learned culture, even though it had some familiar features, such as the use of multiples of quarter-winds instead of the degrees of the geometers. By this time, however, developments of real practical use were beginning, where the techniques of learned geometry were being self-consciously adapted for use at sea.

The challenge of the Atlantic

The uncompromising parameters of physical geography enter our story once again. The Portuguese were well placed to appropriate the navigational practices of the Mediterranean, but their main seaboard offered different prospects, facing outward to the Atlantic. They came later than others to the practice of hydrography, or the making of sea charts, but they were certainly producing portolan charts in the mid-15th century, at a time when maritime exploration was encouraged and sponsored by Prince Henry, later (but not in his own time) called 'the Navigator', with a particular interest in the coast of Africa. That this would lead to new techniques of navigation should not be surprising, after we have seen that circumstances in the Mediterranean had powerfully influenced the emergence of bearing and distance.

The Portuguese project along the African coast, motivated by trade and later by the search for the fabled Christian kingdom of Prester John, depended on discovering what we might call their *southing* at sea, as at this stage the concept of *latitude* or the use of degrees was probably not involved. Some register, however, was vital for recording and revisiting locations, particularly when it was discovered that sailing out into the Atlantic could facilitate finding favourable westerly winds on either the outward or the homeward voyage.

In different navigational cultures we have seen the height of Polaris above the horizon used as an indication of position north or south, but the earliest recorded use of a quadrant to register such an observation by a mariner was by the Portuguese navigator Diogo Gomes in *c.*1460. Astronomers had long used instruments based on a quarter-circle—the quadrant—for measuring altitudes. The standard portable version had a scale of 90° along the arc, two fixed pinhole sights on the radial edge coinciding with 90°, and a plumb-line hanging from the apex. Holding the instrument vertically, the user would sight a star through the pin-holes, trap the plumb-line on the arc with a thumb, examine the scale and read the altitude. This was simple on land but scarcely possible on a moving ship.

Then there was the problem of Polaris being some 3½° from the celestial Pole in the period, and so following a circle of that radius around the Pole every day, its measured altitude varying by 7°. An early expedient—a necessity if degrees were not being used—was to make the observation only when the Guard Stars of Ursa Minor were in a particular orientation with respect to Polaris. The latitude as such would not be found, but the quadrant could be marked for measured positions of Polaris (with the Guards in the appropriate orientation) for individual ports or islands, and the relevant reading maintained while sailing east or west to landfall. This restricted the use of the observation still further. Not only problematic on ship and preferably made, if possible, by going ashore, it could be done only at particular times.

An even more serious difficulty arose when explorations southward meant that Polaris became too low in the sky to be useful—and south of the Equator, of course, it disappeared. Solutions were found for both these problems, but they imported some of the methods of the astronomers and they imposed the use of measurement and calculation in degrees.

The simpler solution was to the displacement of Polaris from the Pole. The *Regiment of the North Star*, as the English would call it, was a set of corrections to the measured altitude of Polaris for eight positions of the Guards in their daily rotation, which could be judged by their orientation with respect to Polaris. The regiment could take a variety of forms—a table, a diagram, an instrument, or a recited formula giving the number of degrees to be added or subtracted. It is difficult to be precise about its introduction, the earliest written direction dating to 1509 and the first published account of an instrument appearing in 1551, but it seems very likely that this practice predated these accounts.

A solution, at least in theory, to losing the use of Polaris on approaching the Equator, is generally attributed to a deliberate intervention in 1484 by the Portuguese King John II, in the form of a commission of relevant experts. Using the Sun was an obvious possibility to consider and a solar solution would be a relatively trivial proposal for a contemporary astronomer. Translating this to seaboard use was a different matter.

As with all celestial bodies, the height of the Sun in the sky depends on the observer's position on Earth, so it looks promising as a parameter for finding latitude during the day. Of course it depends also on the local time, from sunrise to sunset, but that can be factored out by taking Sun's height only at noon, which can be identified because it is then at its maximum for the day. Once the observer perceives the need to reduce the measurement, the noon value has been found. There is, however, a further complication, which we experience as seasonal change: the Sun's altitude is different at the same time, in the same location, at different times of the year. While the stars always maintain the same distance from the Pole and, it follows, the same distance from the celestial Equator, the Sun has an annual cycle of, in astronomical terminology, *declination* or angle above or below the Equator. Solar declination is zero at the equinoxes, +23½° at the northern hemisphere's summer solstice and −23½° at the winter solstice.

At noon, all the angles involved in the relationship are in the same plane: these are the observer's latitude, the measurement of the Sun's altitude, and the solar declination. This makes it much easier to perform the necessary calculations, which would otherwise involve complex spherical trigonometry. We recall that the altitude of the Pole is equal to the latitude of the observer. The altitude of the celestial Equator, 90° distant from the Pole, will be the complement of the latitude. At the equinox a noon measurement of the Sun's altitude will be the altitude of the Equator, so the complement of the observer's latitude. At any other date, the Sun will not be on the Equator and the observer must find its declination from tables and apply this angle to the altitude measurement (see Figure 3). The adjustment for declination had to be applied differently depending on the time of year (declination positive or negative), on whether the observer was north or south of the Equator, and on whether the declination was greater or less than

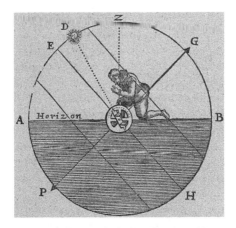

3. A seaman measures the noon altitude of the Sun with a mariner's astrolabe. EH is the equator and D the position of the Sun, in the case shown to the north of the seaman. GB is his southerly latitude. By measuring AD and subtracting the Sun's southerly declination ED, he finds AE, the complement of the latitude. The diagram itself is an evocation the instrument's role in linking the seaman and the geometry of the sphere. W. J. Blaeu, *The Light of Navigation* (Amsterdam, 1622).

the latitude (in the former case the observer would be facing north to take the noon altitude).

Straightforward for an astronomer it may be, but this was a huge step in complexity for the seaman. The declination tables, together with the rules for applying them, became the *Regiment of the Sun*. Either regiment (or rule) enforced the use of degrees, but the solar regiment in particular represents a transformation in the relationship between the art of navigation and the science of geometry. How were degrees to be introduced? How could such measurements be made? Who would supply the tables? The astrolabes of the astronomers each had a degree scale on the back, numbered for altitude, and a centrally pivoted rule with sight vanes (together known as the alidade, mentioned earlier) for aligning it with a star and for reading off the scale. With the instrument suspended by a ring, the scale was oriented with the horizon and the zenith. The astrolabe performed a great many more operations than a simple measurement of altitude, so the appearance in the early 16th century of a more rudimentary adaption appropriate for marine use indicates that there were serious efforts to apply these advanced techniques at sea. We should open a new chapter with this far-reaching development.

Meanwhile, exploration continued to impress the need for new, practical techniques. Bartholomew Diaz, rounding the Cape of Good Hope in 1488, and Vasco da Gama, sailing up the east coast of Africa and crossing to Calicut in 1498, carried astrolabes, but we cannot be sure of their design. That navigators had astronomical instruments, even if they were indeed simplified versions, represents a profound change. Another indication is a report of a new type of chart with latitudes being developed in Portugal about 1485, but none has survived that can be dated to before the 16th century.

Occasionally we can identify specific, personal encounters between practitioners from different navigational cultures. One we have seen already was between a rescued Indian sailor in the Red Sea and the Greek navigator Eudoxus of Cyzicus in 118 BC, which introduced the West to the use of the Monsoon winds for sailing in the Indian Ocean. In 1498 Vasco da Gama, finding Indian traders in Malindi, took aboard a pilot, who guided him to Calicut, and to whom he showed his instruments for altitude. The pilot displayed no surprise, explaining that quadrants were used in the Red Sea for such measurements and wooden tablets of different sizes in the Indian Ocean.

Chapter 3
A mathematical science

The science of finding latitude

Odd as it may seem to someone outside the discipline, historians
of science often have a problem with using the word *science*. This
is because its meaning and reference, as we understand them,
became current only in the mid-19th century. William Whewell is
credited with inventing the word *scientist* in 1833. The historian's
problem, then, is the discomfort of trying often to write the
history of something that did not exist, at least nominally, for
people at the time.

In the case of 'science' (unlike 'scientist') at least the word itself is
not a recent invention, even if its meaning was different in the
distant past. Until the 19th century, 'science' referred to knowledge
of a particular quality or status: it was generalized, systematized,
and assured. It was not the empirical knowledge of the instance,
or even any number of instances, but instances were subsumed
within the principles that formed the science of their discipline.
There might be sciences relevant to all manner of practices—a
science of fencing, for example, or medicine, or preaching—and if
an 'art' referred to the practice of some discipline, its 'science' (if a
discipline had acquired such a thing) was its principles and their
structured connections. It follows from this historical observation
that much modern discussion of the relationship between 'art' and

'science' in the Renaissance, for example, is carried on in anachronistic terms that would make no sense to someone from the period.

The story of navigation is important for the developing relationship between art and science within the discipline of mathematics. In the later 15th century and especially throughout the 16th a number of mathematical authors took up the theme of introducing geometry into a range of practical arts where a potential benefit seemed possible. Geometry could provide shared principles, whose truth was demonstrable, that would underpin a reformed and regularized practice, delivered through the use of mathematical instruments adapted to particular needs. In other words, these disciplines would become mathematical arts underpinned by mathematical science. This had long been the situation in astronomy but one of the first arts where it was shown that such a pattern of development was possible elsewhere was navigation.

We have seen the beginnings of this development in the Portuguese exploration of the coast of Africa and the need to extend astronomical methods for finding latitude. A general rule for finding the altitude of the Pole from the measured altitude of Polaris, and especially finding latitude from the midday altitude of the Sun, obliged navigators to import reckoning angles in degrees from the geometers and instruments for measuring in degrees from the astronomers. It was a far-reaching innovation.

The idea of navigation taking on techniques from other disciplines is reinforced by the appearance of instruments from these disciplines modified for seaborne use. The astronomer's astrolabe was based on the positions of a number of prominent stars and the annual path of the Sun, projected onto a flat star chart (in the form of a fretted brass plate), which could be rotated, on a pivot at the Pole, over a projection of a grid of altitude and azimuth lines between the local horizon and the zenith. A series of plates for the grid would be required to cover a range of latitudes but with these ingredients a variety of astronomical calculations could be performed, with the assistance of a degree scale and alidade on the back to measure altitudes, so as to set the instrument for the current state of the sky. All the navigator would need for his purpose was the altitude measurement, so the *mariner's astrolabe* or *sea astrolabe* would emerge as both a much simplified instrument, having only the altitude scale, which might cover only 90° or 180°, and a much more robustly made one, able to withstand the rigours of use at sea (see Figure 4). Generally a relatively thick and heavy frame was cast in brass in the form of a wheel with vertical and horizontal struts or spokes, crossing at the central pivot for an alidade. This had sights (or *vanes*) closer to the centre than on the astronomer's instrument, with terminal pointers for reading the scale. The closer vanes made sighting easier, as would be necessary at sea, even though it compromised accuracy, while the openwork structure reduced wind resistance. In use the instrument was suspended by a thumb placed through a ring and shackle, and designs with additional weight at the bottom became common, to help maintain the vertical orientation essential to the measurement. An alternative was a wedge-shaped cross-section with more weight at the lower end; the scale on such an instrument would not hang quite vertically, though this may have been offset somewhat by the weight of the alidade.

While the sea astrolabe could be used for observing Polaris, it was particularly suitable for measuring solar altitudes. In the former case it would be held up to the eye and the star sighted directly; in the latter, held at the waist and the light passing through the upper hole was made to fall on the lower. Larger holes were suitable for star sights, smaller could be used for the Sun, and some texts recommended the provision of two pairs, one for each type of observation.

The earliest illustration of a mariner's astrolabe is a tiny marginal sketch of the wheel-type of instrument in a Venetian manuscript

A mathematical science

4. Mariner's astrolabe, possibly Spanish, _c_.1600. In this example only one quadrant is graduated, but the scale is numbered both for altitude (for use with Polaris) and zenith distance (for the Sun). Sea astrolabes were working tools, not collectors' treasures, and paradoxically many have survived by being lost at sea. This one was dredged up in the Gulf of Mexico.

account of Portuguese practice, written in 1517. A larger drawing dated 1529 of an 'Astrolabio Maritimo' on a Spanish chart shows a disc-type, that is, having a solid plate; and the same form is described in Spanish printed navigational manuals of the mid-century.

So there was some variation before the wheel-type became standard. The instrument would become especially common on Spanish and Portuguese ships. The English, French, and Dutch came to it later (with the Dutch introducing some imaginative variants) and moved on to other instruments more quickly than the Iberians. Although the pattern of decline in the use of the sea astrolabe varied between seafaring nations, it characterizes the 17th century as a whole and was largely complete by its close.

The instruments that replaced the astrolabe also replaced brass with wood: the cross-staff, followed by the backstaff. Like the mariner's astrolabe, the seaman's cross-staff (see Figure 5) was a simplification of an instrument used by astronomers. Known variously as the *radius astronomicus* or *baculus astronomicus*, it was familiar to mathematical astronomers, such as Johannes Regiomontanus, Johann Werner, and Peter Apian. A central aperture in a straight rod (the *transom*) allowed it to slide along a longer *staff* or *radius*. The astronomer, with his eye placed at one end of the staff, adjusted the transom until it just covered the angle to be measured. A scale on the radius might give the angle in degrees or it might be a linear graduation, to be converted into degrees by a table. Variations were possible in the mounting, the near sight, and the two far sights. These astronomical instruments could be quite large: Tycho Brahe had one over a metre long, made in Louvain, but Thomas Harriot used one 12 feet in length. The astronomer's instrument might be in brass or wood, and was often mounted on a stand.

The seagoing equivalent was relatively small (some 80 cms long), made of wood, and hand-held. In fact a notable advantage it had over the mariner's astrolabe was being light and, once disassembled, very compact. The staff was square in cross-section, the cross-piece, transom, or vane, shorter (some 40 cms) and flatter. Graduation was directly in degrees. From the late 16th century, scales for different ranges of angles begin to appear on different faces of the

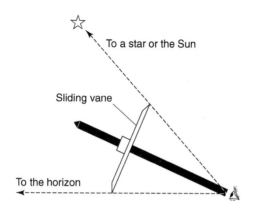

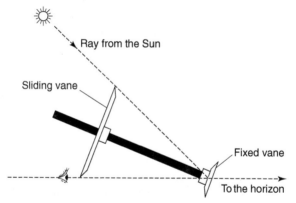

5. **Two ways of using the cross-staff to measure an altitude. In the upper, traditional configuration, the sliding vane is adjusted to cover the angle between the horizon and a star or the Sun, and its position noted on a scale on the long 'staff'. The lower configuration can be used only with the Sun, since the shadow of the upper end of the sliding vane is made to fall on the fixed vane, which with the lower end of the sliding vane is kept aligned with the horizon.**

staff, each to be used with one of a set of vanes of different lengths (the shortest for the smallest angles), but the survival of any original vanes is very rare.

Portuguese and Spanish navigators used the cross-staff in the 16th century, more particularly for the altitude of Polaris, preferring the astrolabe for the Sun; in fact the increasing rarity of the larger holes for star-sights in alidades on astrolabes is attributed to the increasing use of the cross-staff. It was taken up by the French and the English, but it was Dutch seamen who were its chief exponents and it was they who developed the design and introduced new configurations for use. The astronomers used the cross-staff for taking any angle in the sky, but for the seaman it was an altitude instrument. So, the instrument was always to be used in the vertical plane, one end of the transom targeted at the horizon, the other at Polaris or the Sun.

Three problems are quickly apparent. The observer has somehow to sight simultaneously in two directions. The centre of the angle to be measured is at the centre of the observer's eyeball and so inaccessible—some allowance should be made for that, perhaps a numerical correction or by removing part of the staff. The solar sight necessitates looking directly at the Sun: seamen were advised to attach a piece of coloured or smoked glass to the upper end of the transom or, *faute de mieux*, to measure to the top of the Sun's disc and subtract a correction for the semi-diameter.

All of these problems could be removed by adapting the instrument to a *back-sight*, that is, by turning one's back to the Sun. A short vane (the *horizon vane*) should be fixed to the 'normal' eye-end, the eye placed at the lower end of the transom, sighting across the horizon vane to the far horizon from the Sun, and the transom moved until the shadow of its upper end falls on the horizon vane. In this way, the angle subtended at the 'normal' eye-end is, as before, the altitude of the Sun. The Dutch also developed the technique of clamping the transom and adjusting the horizon vane, and in the Netherlands—and only there—cross-staves continued to be made throughout the 18th century. We tend to talk about the fore-sight as the normal configuration of the cross-staff, but since it survived longest among the Dutch, who were using alternative arrangements, this is probably misleading. That it was different from the astronomical configuration indicates that navigators were gaining the confidence to do more than accept modified instruments and to innovate on their own behalf.

Such a trend is even more evident with the introduction of the backstaff, a new instrument specifically designed for finding latitude from the noon altitude of the Sun. It was not adapted from the practice of astronomers but first appeared in a book by a seaman, Captain John Davis, famous for his three Arctic voyages in search of the north-west passage. *The Seaman's Secrets* of 1585 has two designs for the back observation of solar altitude, where, as with the cross-staff, a vane is adjusted along a graduated staff. In the standard design of backstaff that became popular with English seamen in the 17th century and survived through the 18th, the adjustment was made by moving a sight along a graduated arc, but even so, Davis was acknowledged through the popular English name of *Davis quadrant*, though the Dutch and French tended to refer to the instrument as the *English quadrant*.

In the standard pattern (see Figure 6) two arcs together made up the 90° quadrant—an upper, shorter-radius arc of 65° and a lower, longer-radius arc of 25°. They are constructed on the same centre and held in juxtaposition by a standard framework of wooden struts incorporating two holding points for making the observation. There are three vanes, either for sighting or for casting a shadow: one on each of the two arcs (both adjustable, being held friction-tight by springs) and one fixed at their common centre (the horizon vane). The reason for splitting the quadrant in this way is that the observation can be arranged so that the adjustment is made on the larger-radius arc, which, because of its greater length, can be divided with greater accuracy. The instrument would be unwieldy if the longer radius were carried through the whole quadrant.

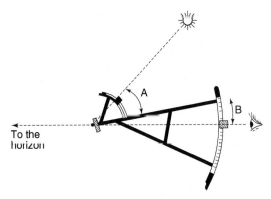

Navigation

6. Use of the backstaff. The shadow of the vane falls on the fore-sight while the sights are trained on the horizon. Adding angles A and B gives the altitude of the Sun.

The upper vane will cast the shadow and first the observer sets it to some 10° or 15° less than the expected measurement. Since the 65° arc is used only for setting the shadow vane to a whole number of degrees, it need not be sub-divided beyond 5°. The instrument is held vertically and pointed to the horizon distant from the Sun. The shadow is kept on the horizon vane, while a view of the horizon is maintained through a hole in the near-sight and a slot in the horizon vane. As noon approaches, the near-sight has to be moved down the 25° arc and, when the navigator would be obliged to begin moving it up again, he stops and takes the altitude reading. Everything is done to make the observation convenient for the purpose: for example, the two graduations are numbered from a shared zero, so that their two readings need only be added.

The 25° arc, with its greater length, imports a scale from astronomy—the *diagonal* or *transversal* scale used, for example, on the measuring instruments designed by Tycho Brahe and by Johannes Hevelius. Diagonal lines are engraved across the

10-minute divisions of the main scale, and the diagonals themselves divided by ten equally spaced concentric arcs, bringing the final subdivision to 1 minute of arc. Given the nature of the observation, the instrument cannot achieve this level of accuracy, but the attempt reflects an ambition to develop the 'science' of navigation, and is the more telling since the backstaff emerged from the community of navigators and instrument makers.

We have noted *latitude sailing* practised in a number of contexts and cultures: it seems a natural response to the ability to gauge position north or south, whether or not conceptualized as latitude, that the seaman should attain the required value and maintain it while sailing east or west to landfall. Instruments such as the astrolabe, cross-staff, and backstaff, systematized this technique and made the concept of latitude in degrees a familiar one.

I should mention one further instrument of nautical astronomy: the nocturnal (see Figure 7), used for finding local time from the rotation of the stars around the celestial Pole. The user would adjust a disc with a time scale against a concentric scale of dates, view the Pole Star through a central hole, and turn an index arm about the common centre until it coincided with certain stars in Ursa Major or Ursa Minor (the initial setting of the time scale being specific to one of these constellations). As the stars rotate around the Pole, the index will indicate the time. Some nocturnals are small and fine, and are clearly intended for use on land, but instruments in a distinctive class of wooden English nocturnals from the 17th and 18th centuries have two additional functions.

One is for calculating the time of high tide, given the age of the Moon and a time constant (the *establishment*) for a port. The establishment, a specific and characteristic constant for each port, is the time interval between the Moon's crossing the meridian at new or full moon and the subsequent high water. (This will be the *apparent*—as distinct from the *mean*—solar time of high water, as

7. An 18th-century English nocturnal. The inscription 'BOTH BEARS' indicates that the time can be found from the positions of stars in either Ursa Major or Ursa Minor.

in both cases the Sun will be on the meridian—noon—at the same time as the Moon, either in conjunction or in opposition.) The other additional function yields a correction to be applied to the measured altitude of the Pole Star, to find the true altitude of the Pole; the Pole Star rotates around the Pole with the constellations whose orientation is registered by the nocturnal. These scales clearly indicate the instrument's intended use at sea, and the latter entails quite a sophisticated notion, by measuring a refinement (adjustment to the true altitude of the Pole) to a measurement (the altitude of Polaris) taken with a different instrument, so as to find the latitude.

Improving the measurement of latitude, in response to the challenges of exploration and navigation, has illustrated how a practical art might become a mathematical art based on a mathematical science. Some of the characteristics of this process would make navigation a convenient exemplar of how such changes might take effect more generally—in surveying, for example, in architecture, or in warfare. Through the application of geometry (the use of degrees), importing some formal astronomical technique (a knowledge of solar declination), by developing instruments, and by codifying procedures in books and manuals, a new demonstrable practice might emerge and begin to grapple with other challenges for finding the way at sea. An example of navigation being used as an exemplary case can be seen in the creation of a mathematical lectureship in London in 1588: in a reference to Francis Drake's circumnavigation of the world, the lecturer Thomas Hood recommended the study of mathematics, 'If you thinke it a blessed thing, to compasse the worlde, and returne again enriched with golde'.

A new projection, a new chart, and new instruments

The new role for mathematics in the changing practice of navigation is nowhere better exemplified than in the development of nautical charts, and the effect this would have on instruments and ways of working at sea. The portolan chart, although a flat representation of an area of a sphere, was not based on any systematic projection. In the confines of the Mediterranean, this was not a great inconvenience; the inconsistencies could effectively be fudged in the collusion between mariner and chart-maker. As ambitions and horizons broadened, however, needs and expectations would change. There were two aspects to this expanding ambition. One was practical, of course: seamen needed more reliable and better-grounded techniques for their longer voyages, particularly over wider ranges of latitude. But intellectual ambitions also were changing. Mathematicians brought to the discipline of navigation the sense that methods and techniques should not only work but should be *right*. In this episode we shall see the spur to be right pressing ahead of the need to be practical, creating what would be a productive tension between the two.

In the case of the portolan chart, it would be immediately evident to a mathematician that a consistent projection was needed to give the chart a legitimate place in a mathematical art. At the same time, any navigator who accepted the visual promise of the compass rose and followed a chosen bearing using his magnetic compass would find that his vessel would not keep to the course indicated by the corresponding line on the chart. This was not simply due to the many difficulties of maintaining a direction at sea; however successful he was at that, he would not make the predicted landfall. The portolan compass rose was a deception. After some disagreement and confusion around the question, the mathematicians concluded that lines of constant bearing, known as *loxodromes*, would be spiral lines on the globe. Whatever shape they took on a flat chart would depend on the properties of the projection used to draw it. The loxodromes intersected the meridians at a constant angle, equivalent to the constant bearing, but, since the meridians themselves converge with increasing latitude, they and the loxodromes would all meet at the Poles.

Before a solution was in place, or before such a solution was understood and widely accepted, charts were needed for areas in size and location not covered by the traditional portolan, which in any case was declining in production and use. A general term, the *plane* (or sometimes *plain*) chart, would come to signify one not constructed on a geometrical projection, but until an alternative was available, this qualifying distinction was not needed. Ptolemy's coordinate system of latitude and longitude, explained in his *Geographia*, could be appropriated, with the addition of scales of degrees, the longitude scale usually placed along the Equator, with or without our familiar sets of parallel lines crossing at right angles. A pattern of radiating rhumb lines remained on sea charts but was joined by lines of latitude and longitude.

On the plane chart longitude lines or meridians were drawn parallel, ignoring the fact that in reality they converge towards the north and south, and the orthogonal latitude lines crossed them at regular intervals. Since meridians are not parallel on the spherical Earth but converge with increasing latitude, a table might accompany such a chart, giving the lengths of degrees measured along different parallels of latitude. The length of a degree is the same for any great circle (one whose plane passes through the Earth's centre), provided the Earth is a perfect sphere, and that applied to every meridian, but the only great circle among the parallels of latitude is the Equator.

An alternative appeared in 1569, and it is significant that its origins are found in what we might call a mathematical workshop—a site of production, where maps, globes, and instruments were made and sold, based on expertise and innovation in practical mathematics, a locus, in other words, of the mathematical arts. This was the workshop in Louvain, established by Gemma Frisius and managed by Gerard Mercator, though by the time his world map was made, religious persecution had caused Mercator to set up his own workshop in Duisburg, in the duchy of Cleves.

Mercator was a cartographer, globe-maker, engraver, printer, and instrument-maker. Diverse as these skills may sound today, in the 16th century they sat together at the core of practical mathematics, where the science of geometry was turned to use in the mathematical arts. The principal geometrical property of Mercator's great map was designed for navigation: all rhumb lines were projected as straight lines on the map. On a chart, this meant that a navigator could set down a rule between two points and find the compass bearing that would take him along the indicated course. Such was the power of this feature that the Mercator projection would come to be used on all sea charts. Why it also became the basis of the most familiar map of the world in many other contexts, including education, is less easily explained.

It follows from the need for straight rhumb lines that meridians will all be parallel on the chart, just as they were on the plane chart. Mercator, however, took on board the implications of such an imposition for other features of his map. The relationship of the distance between the parallel meridians on the chart and between the converging meridians on the globe is constantly changing; in other words, the scale of the chart's east–west coordinate is changing—increasing with latitude. Instead of drawing his parallels at regular intervals, Mercator applied this changing scale to all distances, which resulted in a progressive increase in the spacing of the latitude lines.

We can think of this as a cylindrical projection. The globe sits in a cylinder, touching it around the Equator. Projection lines are drawn from the centre of the globe, through its surface features and onto the cylinder. Once unrolled, the cylinder is the map. Meridians will be parallel vertical lines; lines of latitude will cross the meridians at right angles and their spacing from each other will increase towards the Poles (which cannot themselves be drawn, being infinitely distant). The scale to which such a map is drawn will change with changing latitude. Mercator did not explain the mathematics, but in trigonometrical terms, the scale changes by a factor equal to the reciprocal of the cosine (the secant) of the latitude. This introduces great distortion of areas: in higher latitudes land masses appear much larger than their true size relative to those near the Equator. That does not concern the navigator, provided he or she applies the distance scale correctly, but it does make for an odd view of the world.

The further developments around *Mercator sailing*, that is, the practical application of the Mercator chart at sea, illustrate the new character of navigation as a mathematical art, and also confirm the growing importance of northern European ambitions. The mathematical account, not provided by Mercator himself, was supplied by the English mathematician Edward Wright, in a vernacular publication of 1599, *Certaine Errors in Navigation*, his first target being the errors of the sea chart. Wright explained the geometry of Mercator's projection and provided a table of what are called *meridional parts*, which could be used for constructing such a chart. This was a table of the lengths along a meridian on the Mercator chart, between the Equator and parallels of latitude (at intervals of 10 minutes), expressed in units of 1 minute of arc at the Equator.

Setting a course was one thing, but a navigator adopting the Mercator chart would face a new set of challenges. Finding the distance, for example, of any course involving a change of latitude would, in theory, imply continuously changing scales, though in practice *mid-latitude sailing* could be adopted, which meant using the scale that applied to the middle of a straight course. There was no real prospect of any of this coming into regular practice without the formulation of standard routines to be followed and the invention of instruments for dealing with unfamiliar trigonometrical calculations.

The growing English commitment to practical mathematics is evident in a response that would have outcomes reaching far beyond the original aim of improving navigation. In 1623

Edmund Gunter, Professor of Astronomy in Gresham College, London, published another vernacular work, though with a Latin title, *De Sectore et Radio* (on the sector and radius, meaning, in fact, the cross-staff), where he introduced both instruments and routines for applying them to the calculations required for Mercator sailing.

The first such instrument was a form of *sector* known thereafter as *Gunter's sector*. The sector was a relatively new calculating instrument where pairs of identical scales engraved on the arms of a folding rule, and radiating from the pivot, could be used by surveyors and military engineers. By opening the arms so that the angle between them was such as to set the distance between two corresponding points on the scales to a particular value, the *ratio*, or *scaling factor*, thus established would apply to all other distances between corresponding points. If the scales were linear, this technique—worked with a pair of dividers—has obvious applications in surveying, such as drawing plans to scale. But other lines could be drawn, for handling areas or volumes, or for the weight of shot of different sizes and materials.

Gunter saw the opportunity to extend this technique to navigation, and his sector had lines for the trigonometrical functions required for Mercator sailing. He includes also a scale of meridional parts. The Gunter sector was commonly manufactured throughout the 17th and 18th centuries, but his second instrument, despite its lacklustre name, *Gunter's rule*, continued for much longer. Logarithms were a recent invention and Gunter was one of the first to see their potential for simplifying calculation.

Before moving to Gunther's second instrument for navigational calculation, we might pause to remember (or learn) what logarithms are and why they are so useful in this context. Their immediate practical value derives from the fact that to multiply two numbers expressed as the same number (known as the *base*) raised to a particular power (the power, which is the *logarithm*, being the number of times the base has to be multiplied by itself to yield the given number), the operation is simply that of adding the two powers. So,

$$a^b \times a^c = a^{(b+c)}$$

For division the logarithms are subtracted. The logarithms of numbers (commonly to base 10) can be provided in sets of pre-calculated tables, while tables are also used to convert in the reverse direction, from the logarithmic answer to the number which is the solution required. An alternative to tables is to set out a scale where lengths are proportional to the logarithms of the numbers marked on the scale. In this case the calculations are performed by setting lengths off against each other and the answer is read directly from the scale.

Gunter was so committed to this technique that he calculated logarithmic tables of trigonometrical functions, giving the values according to angles in degrees and minutes, and publishing these in 1620. On his rule, which he first recommended should be engraved on the radius of the cross-staff, and which he published in *De Sectore et Radio*, he set out lines of logarithms of numbers and of trigonometrical functions. With a pair of dividers, lengths could be added together (for multiplication) or subtracted (for division), and it was not long before instrument makers were offering the *sliding Gunter*, where two such scales were mounted in contact and could be moved against each other. Although the static Gunter rule and dividers remained popular with seamen, the *slide-rule*, in its many forms and applications, became the most ubiquitous instrument for calculation until it was displaced by the electronic calculator in the 1970s.

Gunter also set out the routines to be followed with care and deliberation in using both his instruments in navigation. While this made the calculations possible, they still seemed long, complex, unfamiliar, and tiresome. Little wonder that, in comparison with Mercator sailing, *plane sailing* was more straightforward, and *plain sailing* came to refer to something that was easily mastered.

Magnetic variation

On his mission to root out the errors in navigation, Wright moved from the sea chart to the mariner's compass and the issue of magnetic variation. It had become known in the 15th century that the magnetic needle did not point precisely north. Although no compasses from this period survive, there are records of makers off-setting the needle below the card in an attempt to allow for the discrepancy. This practice survived even though it soon became clear that the discrepancy was not constant but changed with location; on occasions differently corrected compasses were carried for different sections of a voyage.

Navigation

The angle between the true meridian and the magnetic meridian indicated by the needle is *variation*, and it became a subject of serious concern and study in the 16th century. In the 1530s Portuguese writers such as Pedro Nunes and Francisco Faleiro (the latter in the service of the Spanish King) describe an instrument for measuring variation—a *variation compass*—incorporating a magnetic needle and a *gnomon* (the component of a sundial which casts the shadow). The meridian is found as the mean direction between two shadows of equal lengths and is compared with the direction of the needle, the difference being read in degrees east or west. Senior pilots, notably João de Castro in Portugal and Jean Rotz in France, collected values and pondered the pattern of variation and how it might be applied to navigation.

Here again, the new English school of practical mathematics had important contributions to make. In 1581 two tracts were published simultaneously in London; they were sold bound together and both were written by men with substantial experience of the sea. One was by the navigator and explorer William Borough, and concerned magnetic variation and its measurement by the variation compass; the other, by Robert Norman, dealt with the newly discovered phenomenon of magnet inclination or dip (the angle a freely suspended magnetic needle makes with the horizontal). Norman was an instrument maker and self-styled *hydrographer*. He and his customers found it an inconvenience that a magnetic needle, balanced at its centre, would not sit level. This led him to investigate the *dip* and to measure it with an instrument he designed, the *dip circle*. Now there were two deviations of the magnetic needle, on whose constancy seamen depended, in the horizontal and in the vertical planes, and both varied according to place.

It was becoming important to find some systematic account of these phenomena and their variations. As William Borough put it, 'I reserue, with intent (if it bee possible) to finde some Hypothesis for the saluying [saving] of this apparant confused irregularity'—an intention characteristic of a mathematical science. A general study was undertaken, by William Gilbert, and the mathematical and navigational content of his influential *De Magnete* of 1600 was supplied by Edward Wright. Matters were, however, to be complicated further. Borough had reported a measurement of variation in 1581, which Gunter checked in 1622 and found a significant discrepancy. When a further change was recorded in 1634 by Gunter's successor as Gresham Professor of Astronomy, Henry Gellibrand, it was clear that variation changed not only with place but also over time.

Variation was important for navigation in at least two respects, which are linked to our two fundamental parameters of direction and position. It compromised, or at least complicated, the use of the steering compass. It became essential, by using a separate variation compass, to keep an eye on the variation. Dutch steering compasses from the 17th century often allow the needle to be adjusted according to the latest measurement of variation, but it was more usual simply to apply a correction to the magnetic bearing being steered. Certainly it became common practice in the best managed ships to take the variation frequently—morning and evening, weather permitting.

The second area where variation became important concerned position finding. This was both aspirational and practical. In the absence of a reliable method for finding longitude, there was considerable interest in using variation for finding position east–west. If variation altered across the globe in some systematic and predictable way, it might be possible to find longitude from simultaneous determinations of the two measurable quantities of latitude and variation and applying the relationship derived from this global account or hypothesis. The Spanish cosmographer Martín Cortés believed this could be done, and the English navigator Sir Humphrey Gilbert agreed. It is interesting that the discovery of secular change in variation did not put an end to such theories; instead it convinced their proponents that the theories had to be more complex, and to take account of the new variable of time.

Others, such as the Dutch mathematician Simon Stevin, held that the global pattern of variation was entirely contingent and could not be predicted by any theory. It could however be discovered by a global survey, or at least one covering the usual sailing routes. Secular changes would then mean that any such *hydrographic* exercise would need to be updated from time to time. By measuring latitude and variation, navigators could find their location in this global pattern and so, in a sense, know where they were, even if they did not in fact know their longitude.

It is often said that position-finding ambitions such as these ended in disappointment and failure, but that is true only of the global ambitions. They did not yield a general method for finding longitude but variation was used in local contexts. The frequent recording of readings to regulate the steering compass accumulated a great deal of knowledge about variation. This may not appear in systematizing textbooks, but logs kept on voyages show that navigators would use a local knowledge of variation to decide whether they were east or west of an island, or how far they might be from a landfall to the east or west. They did, in a sense, use the method proposed by Stevin, but it had arisen through practice and experience, not through global organization.

The changing culture of navigation

In Chapters 1 and 2 it was appropriate to treat regional navigational cultures as, to a large extent, separate entities, each bounded by a combination of geography, ethnicity, or in some sense nationality, combined with distinctive sets of social ambitions around trade, war, empire, learning, or piety. Where they had elements in common, these were derived from the shared characteristics of the sea and the sky. We have now reached a point where that feature of the story is changing. In places, such as the South Pacific, navigational practice did remain, for the present, untouched by developments elsewhere. Individual maritime powers would continue to have strengths and weaknesses in navigation and hydrography, while the seafaring and maritime trading of some peoples would thrive, as others would decline. But a range of organizational and social factors would begin to create a more international culture of navigation, and in particular where there was innovation, invention, or technical improvement, it will not be possible to continue with parallel accounts.

There were a number of reasons for this change. One was the creation of institutions for developing or teaching the science of navigation. Historians are more cautious than they were about the significance of a Portuguese 'school' founded at Sagres by Henry the Navigator, but we have noted the importance of the commission called together by King John II. From the early 16th century an institution known as the Armazém da Guiné e Indias had charge of the more technical matters connected to maritime trade, including instruments, cartography, and the training of pilots. In Spain the Casa de Contratación, located in Seville, had a great many responsibilities in respect of the maritime empire but among them were navigation and hydrography. Even if institutions were meant to operate under conditions of secrecy, the activity of documenting and organizing information made it more likely for knowledge to be communicated and published, and for technical expertise to be traded and to migrate.

The mathematician Pedro Nunes may have been Royal Cosmographer in Lisbon but his publications taught the geometry of the 'new' navigation to all of Europe. The navigator John Cabot was a citizen of Venice, where he began to sail and trade, before working in practical mathematics in Valencia, and then beginning a career of maritime exploration based in Bristol. His son, Sebastian became a cartographer to the Spanish King Ferdinand in 1512 and rose to pilot major at the Casa de Contratación, before he and his expertise were lured back to England by offers from a Privy Council keen to see England develop as a maritime power. Jean Rotz, whose father was Scottish, left Dieppe for an appointment as Hydrographer to the English King Henry VIII in 1542, returning five years later to a successful maritime career in France. Mathematical science was a portable skill and it was the nature of navigation to engender international exchange.

Other types of institutions were founded with navigational science on their agendas. The Professor of Astronomy in Gresham College was expected to cover navigation in his public lectures and, while this may not always have been followed in practice, the College did become a centre for navigational science. Later in the century a 'Mathematical School' was founded at Christ's Hospital, London, to train boys for sea service. In the Netherlands the cartographer Petrus Plancius founded an influential navigational school. Some of the great maritime trading companies, although not governmental organizations, were sufficiently large and centralized to become repositories of information and expertise, and were keen to improve the technical capabilities of their officers. They were, in effect, part of the institutionalization of navigational science, and Plancius was a founder member of one of the greatest of these, the Dutch East India Company.

The early history of the Company, known as the *VOC* (Vereenigde Oost-Indische Compagnie) illustrates the difficulty of maintaining cultural boundaries in such an era of maritime expansion. The Dutch merchant Jan Huyghen van Linschoten used his position as Secretary to the Portuguese Viceroy in Goa secretly to copy many Portuguese charts and sailing directions, which he published in 1595. These were vital to the subsequent success of the VOC, and also of its rival, the English East India Company, which became, in effect, another institution of navigational learning and development within a corporation.

Instruments were themselves mediators of navigational techniques. They could be freely traded, or they could be captured with ships and sailors. Charts were also commodities of exchange, by fair means or foul, and they could transfer navigational science as well as cartographic knowledge. This became particularly so once printed charts became common, and there is a link here with instruments and their makers. Much of the skill of instrument making at the time involved engraving on brass, and a number of the best makers were able to engrave in reverse on copper for printing maps and charts, as well as paper instruments.

Books were the most direct and effective agents for sustaining a widely distributed navigational culture. Pedro de Medina was pilot-major at the Casa de Contratación in the mid-16th century and his practical and influential textbook, *Arte de navegar*, was translated into Italian, French, Dutch, and English. Copies were taken to sea by Frobisher and by Drake. A work by another instructor of the Spanish pilots, *Breve compendio de la sphera y de la arte de navegar* (1551) by Martín Cortés, was translated into

English as *The Arte of Navigation* (1561) and spurred William Bourne to prepare the first English manual, *A Regiment for the Sea*, published in 1574. This too was translated into Dutch.

The first published nautical atlas was the work of a Dutch cartographer, Lucas Janszoon Waghenaer, whose *Spieghel der Zeevaerdt* appeared in 1584. He had been a seaman and had come into contact with Portuguese, Spanish, and Italian navigators and their charts. He joined a traditional set of coastal sailing directions and profile views with an atlas of charts. Enormously influential, his work was translated into English, French, German, and Latin. The English had adopted the word *rutter* from the French *routier*, to refer to a set of sailing directions; now they took to calling a set of charts a *waggoner*.

I have already mentioned the influential mathematical works of Wright and Gunter, and many derivative textbooks were published in the wake of these pioneers. These cultural changes went together with the development of a mathematical science of navigation and were characteristic of a broader movement encompassing other areas where practical disciplines were becoming mathematical arts.

Chapter 4
Dead reckoning, longitude, and time

Latitude sailing and dead reckoning

What was the state of the mathematical science of navigation at the end of the 17th century? Latitude could be found, weather permitting, using instruments adapted from astronomy and some basic geometry applied to an elementary knowledge of the heavens. A geometrical projection for charts accommodated to the needs of seamen was known, even if not always used, and the calculations it imposed on plotting positions and courses had been explained and made accessible through the design of new instruments. The vagaries of the magnetic compass were better appreciated, even if they could not be predicted in a reliable manner but had to be discovered empirically. A method for finding longitude seemed almost as far away as ever; this remained the 'missing link' that could complete the navigator's art.

We might qualify this by saying that, of course, the longitude was 'found' every day, by the routine of keeping a record of the ship's way and extrapolating from the position previously determined. Was this finding the longitude or merely 'keeping' it, since, unlike latitude, once lost there was no way of finding it again? Yet in the absence of a way to secure an independent longitude *fix*, much of the technique for keeping longitude could be managed mathematically, through a combination of latitude sailing and dead reckoning. The latter involved setting a course on the chart, which, if it was drawn on Mercator's projection, would immediately yield a compass bearing the seaman should follow while he estimated or measured speed and recorded the passage of time. This would yield a distance travelled, used to infer latitude and longitude, so as to know where to *prick* (literally, by making a small indentation or a tiny hole) the new position on the chart, but this inference was complicated by the need to take account of the convergence of the meridians on the globe or the equivalent extension of the separation of the parallels on the chart.

In addition to the chart, a range of instruments came to be involved in the routine of dead reckoning. We have already encountered the steering compass several times in the general form that it would retain, namely with the needle hidden below a card bearing the familiar compass rose. The rose appeared also on the *traverse board*, introduced in the 16th century for recording courses steered without the challenge of taking up pen and paper. The *traverse* was the course followed by the ship, or what the seaman called the ship's *way*. A wooden disc with a handle was painted, or sometimes scratched or carved, with the rose design; it had thirty-two sets of eight holes arranged in straight lines radiating from the centre and representing the rhumb lines of the compass. Beginning at the centre, a peg would be placed in the first circle of holes, to mark the heading steered for a half-hour, timed by the running of a sand-glass. The glass was turned and a peg placed in the second circle would mark the subsequent or continuing heading, and so on. When the outer circle was reached, the board contained a record of headings steered over a four-hour *watch*.

So much for direction, what about distance? Since duration was being measured by the sand-glass, speed became the key parameter. This might simply be estimated, or could be measured by throwing a float overboard towards the bow and timing (perhaps by an incantation) the passage of a marked distance on

Navigation

the ship; this is the so-called *Dutchman's log* we encountered in early Chinese navigation. A more reliable measurement came with a different *log*, first described in 1574, in Bourne's *Regiment for the Sea*. A piece of wood in the shape of a quadrant, with the curved edge weighted to ensure it floated upright, was attached by three cords from its corners to a long rope, knotted at regular intervals and wound on to a hand-held reel. It was cast astern and, once it had cleared the immediate wash of the ship, a seaman would count the number of knots played out in a timed interval of thirty seconds. This number was the ship's speed in *knots*, or nautical miles per hour.

The log-line was knotted at intervals of 7 fathoms but, as the nautical mile came to be recognized as the length of a minute of latitude along a meridian, the correct interval depended on the current estimate of the size of the Earth. In the 17th century, mathematicians recommended lengthening the interval, but seamen were slow to change their practice, so it is common to find the small sand-glasses used for timing (the *log-glass*) running through more quickly, often in twenty-eight seconds.

Traverse boards came to include rows of holes beneath the rose for recording measured speeds, perhaps four times in the watch. The board was completed by the helmsman and used by the master in working out a position. The inherent inaccuracies in these crude measurements were not the only obstacles to a successful inference. Account had to be taken of variation, current, and *leeway*. Magnetic variation could itself be measured, but was changeable. Currents were underlying movements of the sea; they might be known from experience or could be sensed from the behaviour of the sea and the ship. Casting the log would at best indicate the ship's speed with respect to the water, but currents could render that very different from speed with reference to the seabed. Leeway was the effect of the wind pushing the ship in a direction different from the heading indicated by the compass. Rules of thumb were used for this correction, or it could be registered by trailing a weighted line astern, by noting the direction taken by the log-line, or, perhaps most commonly, by observing the angle of the ship's wake. All of these adjustments added further inaccuracies.

One measurement was more secure: weather permitting, latitude could be taken from the altitude of Polaris or the meridian altitude of the Sun. If this latitude differed from that of the inferred position, the former took precedence, but the adjustment required to accommodate to the measured latitude could be a revision of either the direction or the distance, or some combination of both. The choice, of course, materially affected the final inference of position. This uncertainty, to be addressed by the experience and judgement of the captain or master in deciding where to prick his chart, usually done daily at noon, arose, of course, from the continuing lack of a direct method for finding longitude.

One further avenue to improving the accuracy of dead reckoning was available through mathematics, if seamen paid a conscientious regard to the simple fact that their traverses took place on a sphere. The fact itself was simple and its truth well understood, but the practical means of dealing with it could be perplexing. Different techniques of *sailing*—plane sailing, Mercator sailing, middle-latitude sailing, great-circle sailing, and so on—entered the navigational vocabulary, each having a different routine of calculation, each with its advantages and challenges.

In deciding the *way made good* (i.e. the actual course achieved and the resulting position), the seaman had either two or three parameters, as well as a starting position from the previous day. These were a bearing (or a series of altered bearings), a distance, and possibly a recent latitude. Lines representing the initial meridian, the course, and the final parallel gave him a right-angled triangle (or a series of triangles). Treated as a plane figure (plane sailing) and given one angle (the bearing) and one side (the distance), this triangle was readily solved graphically or instrumentally, yielding results for progress north or south and east or west. A latitude fix would impose a north-south position and, if this was different from the calculated position, we have noted that a judgement would be required in adjusting to the measured latitude, so that uncertainty was reduced but not eliminated.

As well as the possibility of a graphical solution on paper, some authors promoted instruments in the form of a board (sometimes also called a 'traverse board') incorporating a compass rose, rules, and a square. A more common option was a set of *traverse tables*, setting out the distances required to change the latitude by one degree for the successive points of the compass, together with the *departure* (distance travelled east or west) in each case.

To add to the contingent inaccuracies of the method, there was the absolute inaccuracy that no such plane figure exists on the sea and that the triangle in reality is spherical. Unless the seaman were using a globe for his calculations, which was not done for reasons of practicality and expense, he should operate within the mathematical rules of a plane projection, and Mercator sailing offered him this possibility. In Mercator sailing the distance to be marked on the triangle for a course involving any change of latitude should take account of the changing scale we have encountered through the table of *meridional parts* used for drawing the Mercator chart, which varies as the secant of the latitude. The equivalent relationship for the length of a degree of longitude along a parallel of latitude is the length of a degree at the Equator times the cosine of the latitude. In *Mercator's sailing*, as it came to be referred to in the textbooks (i.e. written with the apostrophe 's'), the procedures for using the table of meridional parts, or those for the routines proper to Gunter's sector or scale, allowed the seaman to apply this secant function in finding a new position from the bearing and measured distance of his course. A convenient approximation for a traverse involving a change of latitude was an adaptation of plane sailing employing the functional values for the mean latitude, in the technique known as *middle-latitude* or *mid-latitude* sailing.

Without suggesting that we have exhausted the techniques and procedures associated with setting and reckoning a course in the 17th and 18th centuries, we should at least mention one further category, namely *great-circle sailing*, if only for the theoretical interest it aroused. Seeking to follow a rhumb line was a simple and practical ambition, but it did not offer the shortest distance between two locations on the Earth. The great-circle route is the arc between two points in the plane containing the centre of the globe. That this is the shortest path was an item of cosmographical interest from at least the 16th century, while explanations of how it might be followed approximately using a series of straight rhumbs plotted on a Mercator chart were in print from the 17th. It was not only mathematical complexity that hampered much practical application: the need to seek favourable winds and the relatively minimal economies achieved between the tropics are only two of the concrete factors that inhibited much use of great circles before the 19th century.

At a practical level, the relative convergence of the format of the ship's log or journal was of great value to the kind of disciplined navigational practice that was essential to successful dead reckoning. Whatever mathematical procedures were subsequently deployed on the record, it was entered for the twenty-four hours from noon into a volume of ruled columns for the hours ('H'), speed ('K'), depth ('F'), compass bearings steered ('Courses'), 'Winds', often 'Leeway', and 'Remarks' on the weather (wind strength in particular), currents, or any notable event or sighting. A number of such log-books would be kept by different individuals on board, with details copied from the original record of times and courses steered, and often independent conclusions were drawn as to the ship's position. The captain's journal had additional columns, for measurements of latitude, and latitude and longitude by account (i.e. the conclusions from dead reckoning), and perhaps also for measurements of variation or of longitude, when that became possible, though these final two measurements might be entered under 'Remarks'. By the 19th century a *log-board* or *log-slate* with ruled columns for chalked entries, kept near the helm, was replacing the traverse board for the basic record of courses, etc.

All the techniques mentioned here evolved in default of a direct measure of longitude. It is often said that in the era of dead reckoning longitude was unknown, but of course that is misleading. Dead reckoning was a method of finding longitude, one that developed in sophistication and accuracy, known to seamen as 'longitude by account'. Further, it remained important to position-finding long after direct measurement of longitude became possible. While it was used in conjunction with, or as integral to, the direct methods, they were critical to the improvement of navigation and it is to them that we now turn.

The problem of longitude

Two foundations in London would prove pivotal to finding two further solutions to the problem of longitude. These were the Royal Observatory, founded in 1675, and the Board of Longitude, which followed in 1714. Each institution was crucial to the emergence of both longitude methods, by chronometers and by lunar distances (or *lunars*). These are generally thought of as different kinds of solution from inferring the longitude by dead reckoning. The distinction is justified, for while both are similarly dependent on observation, record, and calculation, their key measurements are of variables that depend directly on difference in longitude. Dead reckoning calculated the consequences of certain actions, some of them amenable to regulation and observation; each of the new methods made deductions, however complex and challenging, from measurements that depended immediately on the position of the ship.

Both new methods relied on the equivalence between longitude and local time: for every 15° of displacement east or west, local time will change by one hour. If, then, we find the difference between local time and time at a standard meridian, we have the longitude from that meridian. The chronometer method proposed keeping the standard time by carrying a sea-going clock; the lunar method suggested measuring the position of the Moon, which moves sufficiently quickly with respect to the stars or the Sun to act as a celestial clock. To read this clock would require tables of position calculated for time at the standard meridian.

The Royal Observatory at Greenwich emerged as a consequence of the deliberations of a commission set up by Charles II to consider a proposal for a lunar method, and its *mission* was given unequivocally as providing the astronomical basis for this solution for longitude—the positions of the stars and the observational basis for a theory of the lunar motion that would be the foundation for using the Moon as a clock. In time its activities would become entangled also with the chronometer method. The Commissioners for Longitude (they first met as a Board in 1731) were appointed in 1714, in the aftermath of the loss of four ships and some 1,600 men in a single incident off the Isles of Scilly in 1707. An Act of Parliament charged them with rewarding a method that would be 'Practicable and Useful at Sea', stipulated conditions for trials, and stated awards for different levels of proven accuracy. The highest award would be a very substantial £20,000. The Board, like the Observatory, also became involved with both the new methods.

The histories of the two methods are entwined but, since we are to cover technical matters as well as narratives, it will be helpful first to focus more on lunars and then on chronometers, while not separating their stories completely. Neither is the combined story an exclusively British one—and we shall refer in both cases to important and influential developments in France. First, however, we must deal with a vital initiative in instrumentation, the introduction of the principle of reflection in the octant, sextant, and circle. This seems more immediately relevant to lunars but was important too for the full method of using chronometers.

Octant, sextant, and circle

The general principle underlying reflecting instruments is to mount a reflector, usually a mirror but occasionally a prism, close to the pivot of an index arm which moves over an arc with a scale of degrees (see Figure 8). A second sight, often telescopic, is usually fixed to the portable frame of the instrument carrying the

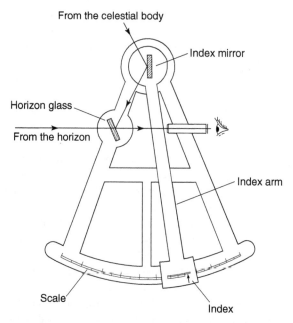

8. Measuring an altitude with the octant. When the image of the celestial body, after two reflections, is brought into coincidence with that of the horizon, the angle between the two mirrors is half the altitude. The same principle is used in the sextant and the reflecting circle.

pivot and the scale, with an alignment such that the user can view one target directly in the line of the fixed sight and another by reflection in the mirror rotated by the index arm. If the instrument has only one reflector (a design seen in early proposals, though no examples survive), the angle moved by the mirror and index to bring the reflected target into coincidence with the directly viewed target is half the angle between the targets (since the reflected ray rotates by twice the angle moved by the mirror). Thus the scale is graduated at twice the rate of the angle it covers (for example, it reaches 90° in a sector of 45°).

The same relationship holds in the common octant or sextant, but now the indirect target is viewed after reflections in two mirrors (which must be parallel when the index arm is at zero on the scale). This has the advantage that the coincidence of the images is unaffected by the motion of the observer standing on deck, provided the instrument is kept pointing in the right direction. The same is true of the reflecting circle, which came into greater favour with French navigators. It follows from the reflecting principle that an octant has a scale graduated to 90°, a sextant to 120°, and a circle to 720°.

There were early designs by 17th-century mathematicians such as Robert Hooke, Isaac Newton, and Edmond Halley, but the history of the commercial instrument dates from the 1730s, when John Hadley proposed two designs in 1731 that were then published in the *Philosophical Transactions* of the Royal Society, one being close to the arrangement that was commonly adopted. In the standard configuration the frame took the form of a 45° sector, with an index arm moving over the graduated arc and carrying a mirror at the apex. A telescopic sight lay across the frame (a pinhole sight might be found on cheaper instruments), mounted on one radius of the sector and directed towards a fixed mirror, the *horizon glass*, on the other. One target was viewed directly, through an unsilvered half of the horizon glass, the other being brought into coincidence after reflection in the *index mirror* and horizon glass, respectively. Equivalent designs were proposed independently at around the same time, such as one by Thomas Godfrey of Philadelphia that was tried at sea in 1730–1 and communicated to Halley in 1732.

The term 'horizon glass' is a clue to a common application of the octant (also known as a *Hadley quadrant*), namely for finding latitude by measuring the altitude of a star or the noon altitude of the Sun. Filters (often called *shades*), mounted on pivots, could be brought into use—at the index mirror for solar observations; at the horizon glass in the case of a bright horizon. Early designers, such as Hooke, thought instruments of this type might be applicable also to the celestial measurements required for finding longitude by lunars, and Hadley provided for this contingency through an optional arrangement that added exactly 90° to the reading on the scale, so as to accommodate the larger angles sometimes required for lunar work. The more practical expedient, however, was simply to enlarge the arc, to accommodate the 120° of the sextant, or even to provide a full circle. The French mathematician Jean-Charles de Borda published a design for a reflecting circle in 1787, where, in addition to the index arm which moves the mirror, the telescope for the direct sight can be rotated around the centre and be fixed anywhere on the scale. This allows the observation to be 'repeated' a number of times in a single procedure and averaged to reduce errors. This repeating circle become popular with French navigators.

The Board of Longitude took an interest in the application of the reflecting principle to portable instruments for use at sea, not least because they had been sent a design for a circle by the German astronomer Tobias Mayer and had arranged trials of an example commissioned from the leading instrument maker John Bird. This was the kind of enterprise that the Board would increasingly foster: initiatives within their general area of interest that, for the time being, might fall short of a full method for longitude. In this case the naval commander in charge of the trials,

John Campbell, saw the advantages of a compromise device that would deal with the angles required but be lighter than a circle. The Board commissioned a 60° instrument from Bird in 1759. Early examples were large and heavy but the Board's continued interest is seen in their support in 1777 for the instrument maker Jesse Ramsden's machine for mechanical division, which could deliver the necessary accuracy on smaller instruments. The sextant was set to become the iconic instrument of celestial navigation.

Lunar distances

The earliest known longitude determinations made at sea by lunar distances were first performed by the French naval officer Jean-Baptiste d'Après de Mannevillette in 1749, near Cape Verde, and then by the French astronomer Nicolas-Louis de Lacaille, made in collaboration with Mannevillette on his astronomical expedition to the Cape of Good Hope, 1750–4. Lacaille published his method with pre-calculated lunar tables (for one month only, July 1761), and his account was available to the Cambridge astronomer Nevil Maskelyne, chosen by the Royal Society for a voyage to St Helena to observe the transit of Venus in 1761. The lunar method was not part of Maskelyne's instructions from the Society but he took the opportunity of this novel experience at sea, as did his astronomical assistant, the mathematical practitioner and teacher, Robert Waddington. Both had their own octants and soon learnt to use them on the deck of an East Indiaman.

It is misleading to think of a single lunar method, as there would be many variants. I introduced the topic, as historians often do, by saying rather glibly that the position of the Moon would be measured with respect to the stars or the Sun and compared with the position calculated for time at a standard meridian. How this might be done was far from clear. Maskelyne knew of two general approaches to the problem: one by Halley had been published in 1752; the other was by Lacaille. Dead reckoning was involved with either method: both required the use of a *longitude by account*, so the traditional inference of position from nautical book-keeping was as much part of this navigational practice as ever. A watch was also needed but, unlike the chronometer method, it was adjusted to local time; it was used simply for keeping the last determination of time from, for example, a noon sight. Even then, the scrupulous navigator would make an adjustment for a known rate of gaining or losing time.

An outline of Halley's procedure will provide some sense of what is involved with the lunar method, but it is important to say that his proposal was only the first of a growing number of alternative sequences of observations, followed by a much larger number of alternative methods of calculation. This would be a significant mathematical industry for years to come.

Apart from taking an altitude to correct the watch, only a single observation is required for Halley's method, which is the Moon's distance either from a star at a similar altitude, or from the Sun, if the Moon is in its first or last quarter. Assume the distance is taken between the Moon and a star. The ship's longitude by account will give an approximation to the time at the standard meridian (which Halley takes to be Greenwich). Using this assumed time, the celestial latitude and longitude of the Moon (coordinates based on the ecliptic) are calculated using the relevant tables and equations, transposed into the Moon's polar distance and right ascension (coordinates based on the celestial Equator), and then the Moon's zenith distance and azimuth (based on the local horizon, i.e. the coordinate system used for the observation). This will give the position of the Moon for the ship's location by account. The zenith distance and azimuth of the star can be found from its tabulated right ascension and declination, applying adjustments for the calculated parallax and refraction as appropriate to these two positions (not the parallax of the star, since that is much too small to be considered). These two adjusted positions yield the calculated apparent lunar distance, to be compared with the measured apparent value. If the navigator is very fortunate or exceptionally skilled in dead reckoning, these values are the same and the longitude by account is confirmed. It is very much more likely that he finds a discrepancy, uses that to make a second estimate of the time difference from Greenwich, and repeats the calculation. This is followed by a proportional calculation using the two discrepancies to find the time difference from the standard meridian, and so the position where the discrepancy would be zero—that is the longitude of the ship.

I have not mentioned a number of instrumental and observational corrections, and, more importantly, I have said nothing about how all these astronomical calculations are to be performed. They require a good grounding in mathematical astronomy and Waddington says that the whole procedure requires some six hours to complete.

Maskelyne began by adopting Halley's procedure but during the voyage to St Helena later shifted to that of Lacaille. This already created an improvement in efficiency: there were more observations but less calculation and, as Maskelyne became more familiar with the officers and their work, he was able to recruit help with observing. In Lacaille's method three observations were required, since the altitudes of the Moon and star were taken in addition to the distance between them.

Both Maskelyne and Waddington realized that a number of challenges lay ahead before the lunar method could become a practical reality. Officers had to be trained in the procedure, and both mathematicians gained experience of this on their separate voyages home. It was important to show that the method did not require an exclusive level of skill but could be readily mastered—an issue that would also become critical for the chronometer method. Maskelyne cited his experience of the homeward voyage in asserting that he did not 'arrogate any particular merit or skill to myself in making these observations, which others may not equally attain with the same care and experience.' Handbooks had to be written and published. Dedicated lunar distance tables would have to be calculated and published well in advance of their dates of use, so as to remove the bulk of the calculations required of those using more general astronomical ephemerides and equations. Improvements and efficiencies would have to be devised to make routine calculations less onerous.

Waddington and Maskelyne set about this enterprise separately, the former publishing first with his handbook, *A Practical Method for Finding the Longitude and Latitude of a Ship at Sea* (1763), and then his lunar tables for 1764 in the first (and only) *Supplement*. Maskelyne published *The British Mariner's Guide* in 1763 and the first volume of the *Nautical Almanac* in 1766 for 1767. It has appeared annually ever since. Whereas Waddington had to rely on his commercial enterprise, Maskelyne had become Astronomer Royal in 1765 and the *Nautical Almanac* was published 'by order of the Commissioners of Longitude', of whom Maskelyne was now *ex officio* a member. Both these new positions—Astronomer Royal and Commissioner—would bring him into a working relationship also with the development of longitude timekeepers.

The chronometer method

We can turn from the relatively arcane technical notions that must be in play before lunar distances would occur to anyone as a longitude method, to a much simpler idea: carry a portable timekeeper set to standard time, which could simply be time at the port of departure. A word about terminology: although 'the chronometer method' is widely used to refer to finding longitude from a portable timepiece, horologists are unhappy with a reference to any individual instrument as a 'chronometer' before the term was used in the late 18th century to refer to a special class of accurate, portable timekeepers.

The earliest, recognized, practical project to find longitude at sea by a timekeeper involved an unlikely partnership between a Scottish laird and a Dutch mathematician. Alexander Bruce, Earl of Kincardine, designed a longitude clock in 1661 and, after discussions with Christiaan Huygens, two spring-driven examples were made by the clockmaker Severyn Oosterwyck of The Hague. These and later clocks commissioned by either Bruce or Huygens were subject to sea trials, with limited success. The notable contribution from Huygens was the mathematical discovery and demonstration that the isochronous curve to be followed by a pendulum bob (one that would beat the same time whatever its displacement or *amplitude*) was not circular but cycloidal. (A cycloidal curve is traced by a point on the circumference of a circle rolling along a straight line.) Huygens also discovered how to confine a bob to a cycloidal path, by suspending the pendulum by threads between cycloidal guides or *cheeks*. The common pendulum, swinging in a circular arc, could keep good time if its amplitude was steady, which was unlikely to happen at sea. It would be typical of the longitude quest to lead to significant developments in both the practical and scientific aspects of mechanical horology.

Navigation

The 1714 Longitude Act and the subsequent activities of the Board of Longitude would be linked to many developments in horology in the 18th century. The chief claimant for the award at the Board's disposal would be John Harrison, an unconventional maker, who had not come through the apprenticeship system but had begun work as a carpenter in rural Lincolnshire and arrived in London in the 1730s with a design for a longitude timekeeper. He was encouraged first by the horologist and instrument maker George Graham and then by a group of other fellows of the Royal Society, who in 1735 provided him with a certificate proposing a sea trial for his first marine timekeeper, an initiative that occasioned the first meeting of the Longitude Commissioners as a Board.

We have seen that the Board were willing to support promising developments that did not yet represent a full solution to the longitude problem and over the following three decades Harrison was the chief beneficiary of this policy. He worked successively on three large timekeepers, the second completed in 1739 but the third (begun in 1741) not until 1760, reporting from time to time to meetings of the Board and being given a series of grants that amounted to £4,000. By this time the prospects for a successful outcome to this support must have seemed bleak: to represent a practical solution for longitude, a timekeeper would have to be capable of manufacture in a timely and relatively inexpensive manner by a reasonably large range of skilled watchmakers.

In the end Harrison asked for a different timepiece to be tested in 1761, not the one the Board had been hearing about and supporting, namely a large watch that looked quite unlike his very large and heavy machines. By this stage their financial commitment had been such that a full trial was inevitable. With advice on procedure from the Royal Society, a voyage to Jamaica was arranged, with the watch in the care of Harrison's son William. Several fundamental problems, including inadequate knowledge of the longitude of Port Royal in Jamaica and an inadequate procedure for declaring and applying a *rate* to the going of the watch, meant that a new trial had to be arranged. A chronometer need not keep perfect time, so long as its rate of gaining or losing is known and can be applied as a correction. One condition of the new trial was that any such rate had to be declared in advance.

The wrangling between Harrison and the Board over arrangements for the new voyage, this time to Barbados, are too involved and protracted to be of concern to us here, but one point of relevance is that two astronomers were to go to Barbados, so as to measure the local time accurately by the method of equal altitudes and to find the longitude on land by timing eclipses of Jupiter's satellites. On their voyage they were to test the method of lunar distances, using lunar tables from Tobias Mayer. One of these astronomers was Maskelyne, not yet Astronomer Royal but with his reputation enhanced by the qualified success of the expedition to St Helena.

The upshot of the second voyage led the Board into further difficult issues. Everyone agreed that the watch had kept time to within the limits that would, at least on this basis, entitle Harrison to the full award of £20,000. The problem for the Board was that the Act stipulated not only the terms for the test but also that the award was to be made for a longitude method that was 'Practicable and Useful at Sea'. The inadequacy that now became evident lay in the legislation itself. How could the performance of a watch on a single voyage demonstrate the practicality of a longitude method? Yet the Act seemed to promise the reward separately on each of these conditions. Many questions seemed to challenge confidence in the method, notwithstanding the success of the trial. What horological principles or manufacturing procedures constituted the superiority of this watch? Could they be revealed, explained, formulated, and communicated? Could such watches be manufactured in numbers, in a reasonable time, at a reasonable cost, by moderately competent makers? Had the success of Harrison's watch been a matter of chance in a single instance? Had it depended on the achievement of a wholly exceptional, individual talent? Either of these explanations would be fatal to a general method.

The Board decided to separate the components of the legislation: they would grant Harrison half the full reward, once he had explained the watch to an appointed group of experts, and retain the other half against the time when it could be proved—by means they stipulated—that the watch could go into routine production. They sought new legislation to clarify the matter, and this was achieved in 1765, with the passing of 'An Act for rendering more effectual an Act, made in the 12th year of Queen Anne' (i.e. the original Longitude Act of 1714). Harrison would first be obliged to 'discover', as it was said, his watch, that is, literally to remove the cover and reveal how it worked. A positive report from this procedure resulted in the payment of £10,000. Harrison never was awarded the second half of the prize, but appealed past the Board to Parliament and the King, for what he regarded as his due. This resulted in a further award of £8,750, which had not been approved by the Board, bringing the total (with the £4,000 already granted in support) to £22,750, though William continued to grumble that they had been short-changed.

For all his very considerable ability and his extraordinary tenacity, John Harrison did not solve the longitude problem. His individual story has proved compelling in recent years, but it does not represent a definitive contribution to the broader history of navigation. He had shown that a timekeeper could be successful over long distances at sea, but his watch did not go into general production, as a solution would have required, and its horological ingredients were not reconfigured into a successful chronometer. The principle of rapid oscillation has been identified as an essential insight, but Harrison's means of applying that to a successful sea-borne timekeeper were not copied for long. For many of those ingredients we have to look to other makers.

A successful chronometer has to be made to high technical and material standards, but in addition there are specific matters of principle that need to be resolved (see Figure 9). Some of these are fairly straightforward: for example, to minimize the effect of the ship's motion, the chronometer, like the steering compass, is carried in a gimbal mount to keep it level. A more challenging matter is the need to keep the oscillator, a spring-regulated balance, as free from mechanical interference as possible, even though its motion has to be maintained by a regular *impulse*. Another matter of principle concerns temperature compensation: a balance spring becomes less elastic with increasing temperature and so the chronometer would lose time without some mechanism to compensate for this and for the reverse effect in cold.

9. Marine chronometer by Thomas Earnshaw, London, *c*.1800. Inside the movement has Earnshaw's spring-detent escapement and temperature compensation balance.

Harrison devised the so-called *grasshopper* escapement, which has the advantage of acting with no friction between the escape wheel and the pallets engaging its teeth, but he did not use this device in the prizewinning timekeeper, which has something closer to the normal *verge* escapement. He also invented the bi-metallic strip, composed of brass and steel joined along its length and bending with changing temperature, and he used this to move a pair of *curb pins* that automatically altered the effective length of the spring. Such strips are used in the form of *compensation balance* that became standard for marine chronometers, but they are incorporated into the balance itself, moving weights towards and away from the centre of motion and so altering the period of oscillation. A *detached* escapement also became standard: here impulse is given to the balance only for an instant at the centre of its oscillation cycle, the escape wheel being locked at other times.

The French maker Pierre Le Roy, whose activity in marine timepieces was mostly around the 1760s, worked with both a detached escapement and a form of compensation that was integral to the balance, as well as making the balance spring isochronal. He had a bitter dispute with Ferdinand Berthoud, who was less original in his concepts but more successful in making sea-going clocks. It was a disputatious discipline. In the next generation of English makers, John Arnold quarrelled with Thomas Earnshaw, both of them able makers, who together (though this would not have been their view) contributed the essential ingredients of the successful chronometer. Arnold designed a form of detached escapement, with a pivoted detent (this being the part that *detains* the escape wheel), and he patented this and, most importantly, the helical balance spring, which would become standard in chronometers. Earnshaw, younger by thirteen years, designed the *spring-detent* escapement (with the detent mounted on a spring, so it does not require oiling), which itself became standard, and the compensation balance in its common form. By the late 18th century all the essential elements of the marine chronometer were in place; what remained for solving the longitude by this method was for them to be manufactured in numbers and taken to sea.

Chapter 5
The zenith of the mathematical seaman

Chronometers, sextants, or both?

In a pattern of development not unfamiliar to historians of technology, new solutions to the longitude problem, involving the introduction of new instruments and techniques, had unexpected consequences: they led to a revision of the whole approach to the practice of navigation. It was natural, perhaps, that the long search for a more reliable longitude method had consolidated, in the minds of those seeking to complete navigational science, an operational distinction between the two coordinates of latitude and longitude: the former had been found; the latter required a solution. Once this had been achieved, the expectation was that the two parameters would be determined by their separate methods. The chronometer was the chief tool of longitude by mechanical timekeeping, though an octant or sextant was required for the essential ingredient of local time; the sextant had emerged in the context of the lunar method, though it needed, at the very least, a good watch, and the distinction between a *deck watch* and a *pocket chronometer* would become blurred. In fact what happened was that theorists and practitioners found other ways to combine chronometer and sextant. Position would still be stated as a latitude and a longitude but the two need not be found in separate and distinct operations.

Neither of the new longitude methods was adopted quickly and widely at sea. Each had its problems for common practice. The observational and calculational techniques required for lunars had to be learnt and mastered, and the mathematicians themselves were divided over the particular procedures to be followed. While lunars were tedious but seemed trustworthy, provided the long calculations could be done with scrupulous care, chronometers had to earn the seaman's trust and even then were very expensive. While we need to know more about practice at sea, it seems that both methods made inroads in the first half of the 19th century, lunars somewhat in advance of the chronometer. In Britain the East India Company was ahead of the Royal Navy in the adoption of both methods.

One index of the use of chronometers is the installation of prominent *time balls* in ports, raised and dropped daily at a set time, for checking and adjusting chronometers. The first was mounted at the Royal Observatory, Greenwich, in 1833, to be visible to ships in the Thames. As chronometers became trusted and less expensive, lunars declined in use and had all but disappeared by the beginning of the 20th century. In some ways the methods were complementary (for example, lunars could be used to check the chronometer at sea) and both were underpinned by dead reckoning and, since both depended on astronomical measurement, dead reckoning would always be needed in any case of cloud obscuring the sky. But it is from the same period that new methods were developed, where the two coordinates of latitude and longitude were more closely integrated. Then it was the instruments, the sextant and the chronometer, that became more closely associated, rather than the two methods for longitude.

This development would depend on a new engagement with spherical trigonometry, even if procedures were more followed by rote than derived from an appreciation of the underlying mathematics. The process was already underway in the 18th century in the context of finding latitude. The restriction to the

noon sight or to the availability of the Pole Star or another
convenient star on the meridian was a serious constraint and,
since all celestial altitudes always depend on latitude, it seemed
that some more flexible method ought to be possible. Latitude was
specified, for example, by two altitude measurements of the Sun,
knowing the polar distance of the Sun on both occasions and the
time elapsed between them, though a procedure for extracting
it had to be devised, disseminated, and adopted. A good many
solutions to the *double-altitude problem*, as it was known, were
found, and corresponding procedures and accompanying tables
were published—including, for example, by Maskelyne—but, as
usual, it is difficult to infer use at sea from the enthusiasm and
productivity of mathematicians. It is indicative of the subsequent
developments that the term 'double-altitude' referred first to a
method for finding latitude, but later to one to finding both
latitude and longitude, and in fact only then did it become
commonly used at sea.

Navigation

If the Sun predominated in the earlier double-altitude solutions,
stellar measurement would later play a part. A star's altitude
could be taken at different times, or the altitudes of two different
stars measured simultaneously by two observers, in which
case, instead of an elapsed time, the angle between the stars,
subtended at the Pole (their difference in right ascension) could
be applied as a known quantity, found in an ephemeris. If the
observations were made at different times, which of course was
always the case for the Sun, a correction had to be made for any
intervening *run* of the ship. Some procedures required the use of
the latitude by account and a recalculation if the result showed a
wide discrepancy.

An alternative *ex-meridian* method for finding latitude relied
on taking the altitude of a single body when it was out of the
observer's meridian, within limits appropriate to different
variants. This general approach became fairly popular at sea and
it too spawned a variety of procedures and sets of tables. Again,

it was not unusual for the series of steps to require a latitude by account, and a repeated calculation using the first calculated value, if this was far from the assumed latitude. With the availability of chronometers, the ex-meridian method extended its procedures to finding longitude and the double ex-meridian method to finding latitude and longitude.

For fear of giving the impression that taking an altitude was a trivial matter, we should note that, as well as the skill and aptitude involved, the measurement might have to be corrected for *index error* (an error of the instrument), the height of the eye above the sea (known as *dip*), the semi-diameter of the body (to find the position of the centre from a measurement of the edge or *limb*), refraction, and parallax.

Before moving to the more integrative methods, what might be said about lunar distances and chronometers themselves in the 19th century? In the former case, a procedure involving simultaneous measurements by sextant of the distance and the two altitudes became standard, the set of observations being repeated several times, averaging to reduce error, with the times of observation noted by a watch. This was followed by one of a number of procedures for *clearing the distance* as it was termed (that is, making all the required corrections and finding the *true* angle subtended at the centre of the Earth); most often the method was based on that published by Borda, and used logarithmic computation. The *Nautical Almanac* offered tables of distances from the Moon to the Sun, the planets Venus, Mars, Jupiter, and Saturn, and a selection of nine bright stars. Greenwich times were given for whichever bodies were at suitable distances at three-hour intervals, with times for intervening distances being found by interpolation, which itself involved a correction for the variable motion of the Moon.

Chronometers were in fairly common use by the middle of the 19th century, having been introduced first on exploratory voyages and then for far-distant stations. Their fundamental design changed little and development came to focus on such details as improvements to the compensation balance. The basic balance did not achieve complete compensation and various forms of *auxiliary compensation* (often referred to as such on chronometer dials) were added, usually acting only in extremes of heat and cold. In fact the marine chronometer was such a stable technology, and its examples so soundly made (as they had to be to perform adequately), that they could be serviced and repaired as needed, and eventually the declining demand for new replacements became problematic for the suppliers.

The chronometer method for longitude was generally considered in the 'astronomical' sections of textbooks, where they concentrated on finding the angle at the Pole between the observer's meridian and that of the body whose altitude was taken with a sextant, when time was also taken from the chronometer. If the observed body was the Sun, the angle at the Pole was equivalent to *local apparent time*, and an adjustment for the *equation of time* (the difference between time by the Sun and by a clock, where the variations in the solar day are averaged to give a *mean*) would yield *local mean time*. A comparison with *Greenwich Mean Time* on the chronometer would give the longitude. If the observed body was a star, an adjustment had to be made involving the right ascensions of the star and the Sun. The procedure for extracting the angle between the observer's and the measured body's meridians used a value for latitude, taken from the latitude by account.

Spherical triangles and lines of position

One way we can register the growing integration of nautical astronomy is through the ubiquitous references in textbooks to the *PZX* triangle (or spherical triangle, i.e. one whose sides are arcs of great circles), also known as the *navigational triangle* or *astronomical triangle* (see Figure 10). The convention of denoting

Navigation

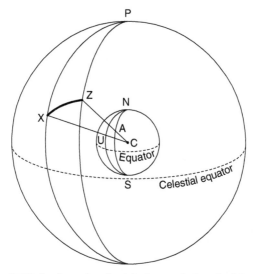

10. **The PZX triangle on the celestial sphere. P is the celestial Pole; Z the zenith point of the observer at A; and X the position of the celestial body. U is the 'geographical position' of the celestial body and C the centre of the Earth.**

the celestial Pole as 'P', the observer's zenith as 'Z' and the celestial body being observed as 'X' became common in explanatory diagrams and proofs even before references to the 'PZX triangle' became pervasive. Many procedures were formulated as solutions to the relevant PZX triangle; the technique just mentioned, for example, for finding longitude by an ex-meridian measure of an altitude, was solving the PZX triangle for the angle P.

In theory a knowledge of spherical trigonometry could be mastered and then applied as appropriate, replacing a situation where long and involved procedures were blindly followed, with little hope of ever understanding how or why they were supposed to work. Even at the end of the 19th century, however, the redoubtable Captain Lecky, whose colourfully entitled textbook *Wrinkles in Practical Navigation* was embraced affectionately by seamen, thought the geometry was still a mystery to most; 'Let us open the pages of this sealed book, by which the navigator may learn to reason for himself, instead of trusting entirely to rules which, when forgotten, leave him adrift on his beam ends.'

We have noted that there were many solutions to the double-altitude problem but one would lead nautical astronomy in a novel and productive direction. Introduced in some books as 'Sumner's method of double altitude', it led to an integrated technique for finding latitude and longitude together. The essence of the idea came to the American sea captain Thomas H. Sumner while at sea in December 1837, under a cloudy sky and in some doubt of his position off the south coast of Ireland. On the basis of one solar altitude he worked out a longitude using his chronometer and his latitude by account. Unsure of the latter, however, caution urged him to try another latitude 10 minutes north, and then a third at a similar interval. Finding the three positions on the chart to lie on a straight line, he realized that his ship lay somewhere on this line and that, by following the course it indicated, he would arrive where the line (the first *position line*) was leading on his Mercator chart. Sumner did not know his position, except that it lay on his position line, but here was an idea that could be developed. As Sumner later explained, if position lines were established from two observations (in effect, marking the calculated longitudes on the assumed lines of latitude) and the first was transferred for the run of the ship during the interval between them, their intersection was the position of the ship at the time of the second observation.

Although straight for practical purposes, these position lines were, in fact, arcs of circles with large radii. An altitude measurement established on the Earth a circle of altitude, or circle of position, whose centre was the point directly beneath the body observed. Following an appropriate lapse of time another altitude could be taken, with the body now at the zenith of a different centre, and the intersections of the two circles (allowing for any intervening run) would yield two points, one of which must be the position of the ship. So, dead reckoning came into this method in two ways: it provided the information for the run of the ship and, by choosing assumed latitudes based on the latitude by account, the position lines would be drawn in the area of the appropriate intersection. If simultaneous altitudes of two stars were taken, the position lines would of course give a fix without allowing for an intervening run.

Other position-line methods were devised. A popular example was published by the French Captain Marcq de Saint-Hilaire in 1875. Also known as the *intercept* method, it involved measuring the altitude (and thus finding the zenith distance) of the Sun, and calculating the Sun's distance from the zenith of the observer's dead reckoning position for the time of observation. The difference between the observed and calculated values is the *intercept* and, if the observed zenith distance is greater or less than the calculated value, the observer's position will be farther from or closer to the geographical position of the Sun (the point whose zenith is the Sun's position) by the intercept value. Accordingly this value is marked off the line joining the dead reckoning position and the Sun's geographical position. Since this line is a radius of the circle of equal zenith distance, the line drawn at right angles to it at the point found from the intercept is a line of position, on which the ship must lie. A second observation is taken later and the procedure repeated, the intersection of the second position line with the first, transferred for the run on the ship, is the ship's position. Once again, dead reckoning is integral to the method. In spite of the longitude problem having been solved, the procedures adopted for effecting the solution at sea draw us back once again to consider the state of dead reckoning.

Dead reckoning

Nautical astronomy had reached new levels of sophistication and versatility in its application to the challenges of navigation at sea.

Yet seamen still needed to follow a bearing and measure their distance. Dead reckoning was, perforce, maintained as a daily routine—in fact more than maintained, it was significantly improved. In the textbooks a standard presentation emerged in the 19th century when three *sailings* were explained: *parallel sailing*, *middle-latitude sailing*, and *Mercator's sailing*. This order was probably adopted for instructional reasons, as a sequence readily explained and understood.

Parallel sailing applied to any traverse due east or west, that is, with no change of latitude, and finding the distance (*departure*) was simply a matter of multiplying the difference in longitude by the cosine of the latitude. At the Equator, the departure in nautical miles is the difference in longitude expressed in minutes of arc, and the cosine function reduces this value as the latitude increases. In an equivalent relationship, the difference in longitude is the departure times the secant of the latitude. When a course follows an oblique rhumb line, that is, one involving a change of latitude, it is often acceptable, in the simplest form of *middle-latitude sailing*, to use the same formulae, while entering the latitude value as the mean between the beginning and end of the traverse. The textbooks generally derive the formulae for finding course and distance in *Mercator's sailing*, which is a rigorous method, without approximation, the tangent of the course being the difference in longitude divided by the separation of the latitudes in meridional parts, and the distance being the difference of latitude times the secant of the course.

Great-circle sailing could be used for shaping a route, but for individual sections a rhumb line had to be followed and calculations made according to one of the methods already mentioned. A great-circle (or *gnomonic*) chart, where great circles are projected as straight lines, could be used to plot the course, and suitably spaced waypoints transferred to a Mercator chart, from which the sectional rhumb lines could be found. The saving in distance could be significant over long voyages in high latitudes.

The *day's work*, as the task was known, consisted of bringing together all the bearings and distances sailed since the previous noon, making allowance for leeway, magnetic variation, and current (resolved into *set*, that is direction, and *drift* or speed, and treated as a course and distance made by the ship), applying the formulae for the relevant sailing, and concluding a new position and a course and distance *made good*. While leeway decreased with the coming of steamships, metal hulls increased the distorting effects of another magnetic variant, an effect local to the ship known as *deviation*.

The marine compass was the object of much criticism and dissatisfaction through the 18th and a good part of the 19th century, but there had been modest improvement. Perhaps it was by contrast with the strides made in astronomical and horological instruments for navigation that the vital tool of the steering compass seemed to have been left lagging behind. By the mid-19th century, however, the standard compass of the Royal Navy, for example, had a copper bowl and four straight needles attached to a ring carrying a paper *card*. The compass rose was a standard black and white to improve visibility. This was a *dry-card* compass and by the end of the century a liquid-filled bowl might be used to dampen the motion of the card and reduce wear on the pivot.

The disturbing effects of iron fittings and objects on the ship were already known before this increased dramatically with the advent of metal hulls. The problem was tackled by creating a binnacle—the housing for the compass—that had a range of features aimed at isolating the magnetic needle from these effects. It is noteworthy that the problem attracted the efforts of eminent scientists in the 19th century—Siméon Denis Poisson, George B. Airy, and William Thomson (Lord Kelvin)—and we might mention particularly the Kelvin binnacle, since it became such a familiar and evocative piece of maritime furniture.

At its most simple level, an iron or ironclad ship became an *induced* magnet simply by being in the Earth's magnetic field and had become a *permanent* magnet as it was being built, through hammering and jolting. The former property varied as the ship moved, the latter was determined by the orientation of the ship under construction. The Kelvin binnacle has two large spheres of soft iron on either side of the magnet, designed to divert the induced magnetic lines of force and shield the needle, while compartments below the compass house a range of permanent magnets set after the ship has been launched to counteract her permanent magnetism and create a null effect at the needle. In addition an adjustable vertical bar of soft iron, attached to the outside of the binnacle—the *Flinder's bar*—corrects for the induced magnetism due to the vertical component of the Earth's field, which varies with latitude. Thomson also introduced and patented a particular design of compass card, in which eight short steel needles are held by silk threads to an aluminium ring with a paper rose. This was designed to minimize friction and wear on the central sapphire bearing moving on an iridium point.

The other major component in the data for dead reckoning—speed—was also made more reliable through improved instrumentation. Mechanical logs, often known as *patent logs*, became a commercial reality at the beginning of the 19th century with a patent granted to the English maker, Edward Massey, for a log where a geared counting mechanism was turned by a pointed cylinder with vanes, the whole apparatus being towed at the stern. This was only the beginning of a succession of designs and patents by Massey and his successors and competitors, the later examples being *taffrail logs*, where the register remained aboard at the stern, while only the rotator trailed in the water. This had the advantage that it could be read without having to be hauled in.

Massey was also involved with the introduction of a successful mechanical sounder attached to the lead. His nephew, Thomas

Walker, who was successful in the development of the mechanical log, also moved into the manufacture of sounders. A different principle was applied in a patent acquired by the American engineer John Ericsson for a sounder where sea water was forced into a glass tube against the pressure of the air inside, to a length that depended on the depth it has reached when finding the bottom. William Thomson applied a similar principle to a successful deep-sea sounding machine, patented in 1876, which could measure down to 150 fathoms with the ship going at full speed.

The coming of age, as it might seem, of mathematical navigation through sophisticated nautical astronomy, was also the period when mechanical instruments raised the bar for some of the most basic and traditional aspects of navigational practice.

Hydrography

If nautical astronomy, chronometers, and dead reckoning were coming together in a shared spectrum of navigational technique, they all depended on the use of the chart. The 19th century saw the standard of state-sponsored hydrographic survey and chart production reach new levels of consistency, reliability, and even cooperation.

The French had led both cartography in general and hydrography specifically through most of the 18th century, founding the Depôt des Cartes et Plans de la Marine in 1720, the setting for the influential fifty-year career of Jacques Nicolas Bellin. Bellin joined the Depôt as a hydrographer to the navy in 1721, becoming the first holder of the office of Ingénieur hydrographe de la Marine in 1741, and under his direction there was a remarkable output of charts, both in number and in standard of production. Publication by a department of state meant, in effect, that this was a navigational service provided for the world. Previous attempts by states to keep hydrographic knowledge secret had failed and commercial enterprises had either produced unreliable charts or collapsed, or both. A common service, however, was consistent with the philosophy of the Encyclopédistes, of whom Bellin was one, with their belief in a universal assembly and dissemination of knowledge.

Some of the same international spirit is seen in the policies of the Board of Longitude, but the British had not the same organizational mastery of hydrography, though the work of the Scottish hydrographer, Murdoch Mackenzie, famous for surveys of his birthplace, the Orkney Islands, led to his proposal of a standard tool of hydrography, the *station pointer*. The idea was taken up by his nephew and successor of the same name, and the instrument came into use in the early 19th century. A circular protractor had three arms extending from the centre, one fixed at zero, those on either side free to move, with clamps to set them to particular readings and perhaps tangent screws for fine adjustment. To locate an offshore position, often of a measured sounding, the two angles between three landmarks were measured with a reflecting instrument, the arms of the station pointer opened to these angles and aligned with the landmarks on the chart, when the centre of the protractor would be at the position of the sounding. Another Scot, Alexander Dalrymple, became Hydrographer to the Admiralty in 1795, having served the East India Company energetically in a similar capacity.

Two outstanding hydrographers led the French and British hydrographic offices in the second quarter of the 19th century, respectively Charles-François Beautemps-Beaupré, in charge in Paris from 1826 to 1854, and Francis Beaufort in London from 1829 to 1855. Beautemps-Beaupré is often referred to as the founder of modern hydrography, for the improvements he made in methodology and instrumentation, including the use of the Borda circle, while Beaufort greatly expanded the coverage of the British surveys and charts, over a thousand being issued during

Navigation

his tenure. Elsewhere there were significant state-sponsored offices in Denmark, Sweden, Spain, and Russia, while the United States Coast Survey issued its first chart in 1835. By the second half of the century British hydrography had taken the international lead from the French, which was the main reason for the choice of Greenwich as the agreed prime meridian of the world in 1884.

Chapter 6
The electronic age

Radio

The origins of a wholly new approach to navigation at sea might be traced to the investigation and application of radio waves, or *wireless* waves, in their early terminology. Research began through interest in a physical phenomenon—of one of a number of 'new' waves in the late 19th century—but once the use of wireless for carrying messages became a realistic ambition, its application at sea was a clear priority. The electric telegraph had transformed global communications but ships at sea remained out of touch, except for direct encounters with other vessels. One of Guglielmo Marconi's early commercial goals for the company he founded in 1897 was to establish a communication system for ships at sea. This was an emphatic success, with the company not only supplying the receiving and transmitting equipment on board but training and employing the wireless operators.

The first radio aid to navigation was a time signal for checking the chronometer, though in fact radio signals would eventually spell the end of the real need for a chronometer, since a good watch could be checked as often as required. *Radio direction finding* (RDF) was introduced at the beginning of the 20th century with radio beacons established to transmit signals, so that a direction-finding antenna could determine a bearing to a known transmitter on shore. The signal contained an identifying code and the procedure was made much more efficient by the development of the Bellini-Tosi finder, patented in 1907. With two transmitters in range, a position fix was possible. I should mention that radio bearings cannot be plotted directly on a Mercator chart, where, as we have seen, it is rhumb lines that are rectilinear projections, while radio waves follow great circles. With smaller receivers and the development of automatic finders, RDF was a successful technology for much of the 20th century.

Hyperbolic radio navigating systems, as they were known, represented a significant technical development in the use of radio for position finding. The term 'hyperbolic' comes from the overlaid lines on special charts issued for use with these systems. A station transmits a pulsed radio signal, which is received on the ship and also by a *slave* station in the land-based network, which transmits a signal of its own after a known delay. This signal is also received at the ship, where the set gives the time in millionths of a second (microseconds) between the arrival of the two signals, which in turn identifies a position line on the chart. There is, in other words, a set of locations where the time difference will have the same value, and they lie on a hyperbolic curve. A second master–slave comparison from a different pairing of stations will give a different value, identifying a second line, which will intersect the first at two points. Another source of information, such as dead reckoning, will allow one intersection to be selected, or a third timing can identify a third hyperbolic line. Ways were found to distinguish different slaves sharing a master, such as by using different frequencies, and of identifying a particular *chain* of master and slaves, where it might be confused with another.

An early example of this technology was the short-distance Decca Navigator System, first used by the Royal Navy in World War II. Here the timing of phases of continuous signals were compared, not discrete pulses. The range was some 500 miles and it was much used in coastal waters, and not finally decommissioned until 2000.

Loran (long-range aids to navigation) was introduced by the US Navy from 1942, later taken over by the US Coast Guard. It was the outcome of extended research into radio navigation sponsored by the US military, latterly carried on at the Massachusetts Institute of Technology (MIT). In this case the transmission was a pulsed radio signal with a range of about 1,500 miles. Loran-C was introduced in 1957, using signals with phase difference and pulse characteristics and a range of up to 3,000 miles. With the longer distances, the signals needed additional features to identify them and avoid confusion. After decades of worldwide use, the system declined with the rise of satellite navigation and the US Coast Guard finally closed its stations in 2010. Elsewhere there has been interest in continuing or enhancing Loran, because of the potential vulnerability of satellite systems, but the UK terminated a prototype Enhanced Loran (eLoran) in 2015.

Other systems could be mentioned, such as Omega (an international system operating between 1971 and 1997) or CHAYKA (a Russian equivalent to Loran-C), but enough has been covered for the present purpose. What should be described briefly is a technology where radio signals are transmitted from the ship, not the shore.

In the technique of *radio detecting and ranging (radar)* a shortwave radio signal emitted at the vessel by a rotating transmitter is reflected by whatever it encounters, whether buoys, other ships, or the coastline, and the returning signals are rendered visible on a screen. Though clearly valuable for safe passage, especially in poor visibility, radar may sound to have limited navigational application, but the distance of the detected objects could be measured and, combined with a gyro compass (which we shall come to shortly), a range and bearing to a known fixed object could be found. By detecting buoys or shoreline features, whose locations were known, what were in effect position lines could be plotted and their intersection found. In 1956 a system manufactured by Decca showed the vessel itself on the screen in relation to the other identified features.

Inertial navigation and dead reckoning

Before introducing satellites into our story, we might switch off the radio for a moment, stop *listening in* to external signals, and consider a contrary and perhaps unexpected development in 20th-century navigation—contrary in that it depended on no external resource in a *live* way and was self-contained to the vessel it guided. The development of the *inertial navigation system* (INS) was mainly driven by the needs of aircraft and missiles, but it was also given some use at sea.

The first element in the repertoire of instruments was the *gyrocompass*. The gyroscope had been of interest in mechanics long before its practical application as a gyrocompass. For example, Léon Foucault, whose interest in spinning bodies is best known from his eponymous pendulum, stated in 1852 that a horizontal axis of a spinning disc tends to align with the axis of the Earth. A freely suspended spinning disc (in a gimbal mount) can be subject to an interfering force that results in it aligning its axis of spin with that of the Earth, that is, it points to true geographical north. In practice many refinements and compensations are needed, for example, to deal with the disturbing effects of a sharp change of course.

The first practical compasses resulted from the work of the German inventor Hermann Anschütz-Kaempfe; they were manufactured by Anschütz & Co. in Kiel from 1905 and under licence in England by Elliott Brothers from 1910. The entry of the Sperry Gyroscope Company into the American market in 1911, and then the British in 1913, led to legal challenges over priority and patent rights, with technical opinion being sought from the sometime patent expert Albert Einstein.

Inertial navigation is a form of dead reckoning and, if direction is one component in the dead reckoning calculation, we know that speed (together with time) is the other. The instrument developed for this purpose was the *accelerometer*, which detects and measures the forces on a suspended mass caused by the movement of the vessel. Newton's laws tell us that nothing will be detected from a steady velocity or speed, but accelerations and decelerations can be measured and an integration over time can compute velocity. Again a vessel is subject to many accelerations that are not directly part of the dead reckoning calculation, so the whole system is very complex, and the theoretical discipline that supports it correspondingly sophisticated. We cannot do justice to it here but, at the same time, very accurate INSs are only relevant at sea to submarines or very large naval ships. Combined with GPS (Global Positioning System), however, which we shall come to shortly, less accurate systems are now in use, where output from inertial navigation can be updated and corrected by satellite navigation.

Traditional dead reckoning continued to be important in the 20th century and indeed to be improved. Patent logs were in use until the 1980s, when they were finally superseded by electrical logs, delivering a continuous display in the wheelhouse. A battery-powered counter in a trailing log could convey its results to the ship via the log line, but it became more popular to mount the log on the hull. In the late 20th century ultrasound logs gave speed relative to the sea bottom, so their readings were not affected by currents.

Echo sounding involves timing the vertical passage of sound emitted from the ship and reflected from the seabed and was developed during and just after World War I. It is a particular application of the general principle underlying *sonar* (i.e. sound navigation and ranging), where sound is sent through the water to detect other vessels and came into use in the 1920s.

Satellites

As the world tuned into the radio signal from the first artificial satellite, Sputnik, launched by the USSR in 1957, scientists at the Johns Hopkins Applied Physics Laboratory (APL) in the USA detected a *Doppler effect*: the signal frequency increased as Sputnik approached an observing station and decreased as it moved away. This was applied in the *Transit* system developed at APL for the US Navy, which first became operational in 1964—the earliest working satellite navigation system. Eventually five satellites could provide global coverage, though at least ten were usually deployed. The profile of frequency change—the *Doppler curve*—was unique to each receiving position, not least on account of the rotation of the Earth, but finding the match was a very complex process and the system did not give real-time results: at best a fix could be found once an hour. Even so, it was widely used, first on Polaris submarines but then for general shipping, including civilian craft. It ceased in 1996, though the satellites continued in service as research tools for monitoring the ionosphere.

The basic notion of satellite-based positioning was so attractive that other proposals were soon to follow. The US Air Force, the US Army, and the Naval Research Laboratory worked on separate initiatives, which they each began to build. In 1973, with the Department of Defense facing development funding applications on three fronts, it was decided that concentrating on a single system would be more effective. The early name 'Navstar' became associated more with the set of satellites, with the whole facility becoming the *Global Positioning System* (GPS). The first of ten prototype satellites was launched in 1978, and the twenty-four that formed the operational complement between 1989 and 1994, the system being declared complete in

1995. More satellites have since been added. Each orbits the Earth every twelve hours.

At first the full capability of the system could be accessed only by the US military, but since May 2000 full accuracy has been available to all users. This will depend on a number of factors, notably the quality of the receiver, but should be between 3 and 15 metres. The possibility of altering the available accuracy is a clue to how the system operates. Each satellite carries two atomic clocks and transmits signals containing coded information on time and orbital data that yield its position, which can be monitored and the information adjusted by ground stations. Control of the coding meant that different levels of service could be provided for different types of user. The GPS receiver can process the signals from four satellites and provide a fix based on comparing the times it has taken for them to arrive.

Meanwhile the USSR built and deployed an alternative, the Global Navigation Satellite System (GLONASS), with twenty-four satellites, each carrying three atomic clocks. The principles are similar to GPS, though the latter's satellites are identified by individual codes, while those of GLONASS use the same identifying code but different frequencies. Development began in 1976, satellites were launched from 1982 and all twenty-four were in place by 1995. As in the US, there was an earlier satellite system, 'Tsiklon', designed for military use, which had been operational since 1972. GLONASS also was made accessible to civilians and receivers that can work in either GPS or GLONASS are now commonly available.

The European Union took the decision in 1998 to develop a system of its own, 'Galileo', now being implemented in collaboration with the European Space Agency. At least twelve satellites have been launched, with the complement of thirty being anticipated by 2020. In addition, China is developing her regional 'BeiDou' system, so as to achieve a global capability by 2020.

An international discipline

As we near the end of our story, we might return to some of the educational, organizational, and societal aspects of navigation at sea that we have left behind since observing, at the close of Chapter 3, that the growth of a mathematical science of navigation and its delivery into practice through the use of instruments was making cultural boundaries less important than before. This feature has continued to grow since the 17th century, underpinned by the developments we have followed in mathematics, physics, and engineering that were at least potentially open to all. By the late 20th century the international character of navigation was no longer grounded mainly in the shared content of books and the common principles of instruments, but was secured even more emphatically by artificial resources built by science and technology, not to mention power and politics, and delivered globally from widely accessible platforms. Differences in practice at sea now had economic drivers rather than cultural ones—they depended on the sizes of individual vessels, their corporate or civil identities and roles, and the geo-political distribution of economic development.

More sophisticated navigational techniques could be adopted at sea only with enhanced training and opportunities increased for writers and teachers in the 18th century. Private tuition, individually or in academies, was an important component, alongside some official provision, which could be initiated by local governments or by commercial corporations, but the balance varied in different settings. In Copenhagen, for example, the Shipmasters' Guild (Københavns Skipperlav) ran a navigational school from 1675. In the Netherlands, the VOC developed a system of qualifying examinations and in 1748 established a joint navigational school with the Admiralty and the city of Amsterdam for training officers. In France, navigation schools in major ports prepared sailors for examination for positions as masters, which

by law required a certificate. German ports, such as Hamburg and
Bremen, also established their own schools. In Britain more
reliance was placed on training at sea, supplemented by a number
of private schools in ports, some commercial, others charitable.

The 19th century saw more widespread, centralized insistence on
training, examination, and certification across much of Europe,
where most maritime nations developed some provision for
centralized, mandatory qualification. Britain maintained an
affection for promotion through the ranks and was unusual in
Europe in not establishing state-run nautical schools. Even in the
21st century, training has been provided at a local government
level, with certification undertaken nationally by the Maritime
and Coastguard Agency. In the USA, the United States Naval
Academy in Annapolis was founded in 1845, while schools for
the merchant service were created by individual states before the
federal Merchant Marine Academy at King's Point, New York,
opened in 1943. Government initiatives regarding merchant
shipping could be seen politically as unwarranted interference
with commercial trade.

Some maritime academies award degrees but navigation is also
now a university subject, though often sitting within more broadly
constructed academic disciplines, such as maritime studies. One
notable example is the World Maritime University in Malmö,
Sweden, established in 1983 by the International Maritime
Organisation (IMO). The IMO is the recognized international
authority for standards of training and individual certification,
under the International Convention on Standards of Training,
Certification and Watchkeeping (STCW), drafted in 1978 and
substantially updated in 1995 (STCW 95). For instance, even
where nautical training in Britain is organized locally, it conforms
to this international convention. To take an example from a
different region, vessels registered in many countries are, for
economic reasons, largely crewed by seamen and sea-women from
the Pacific area, especially the Philippines. Here there are many

nautical schools, where there has been a major effort in the 21st century to meet the standards of STCW 95.

Qualifications issued by institutes of education and training are one route to professional recognition; another is chartered institutes established within professional communities. Although navigation has a long technical history, the foundation of professional bodies came relatively late, compared with disciplines of a similar technical character in science and engineering. The Institute of Navigation, in the USA, began in 1945; the UK Institute of Navigation, in 1947 ('Royal' from 1972). Equivalent institutes were founded in Germany in 1951, France in 1953, Italy in 1959, and so on. Such institutes have all the usual trappings of learned societies, such as research journals, conferences, awards, and grades of membership. The British, French, and German institutes began a series of joint conferences in 1957, which would be held every three years to discuss technical issues in navigation. In 1975 this developed into a permanent association, when the institutes of Australia, France, Germany, Italy, Japan, the United Kingdom, and the United States agreed on the formation of an International Association of Institutes of Navigation (IAIN), with the first General Assembly held in London the same year.

Anyone pursuing a maritime discipline today will very soon encounter an international body of some description, and mention of the IMO and IAIN is a reminder that we must look at this international character if we are to understand how navigation is managed and practised in the modern era. The nature of seafaring is such that issues and problems inevitably arise where it is clear that effective improvement is possible only through international agreement between governments. Yet to an outsider there seems to be a bewildering array of international bodies: there seems to be great diversity in the attempt to find unity. Might this be partly due to a tension between the obvious advantage, amounting to necessity, of collaboration, and the evident competition that has threaded its way through our narrative? The competition to be ahead in the technologies of navigation has often been aimed at trade, exploration, and expansion, ambitions that have also led to hostility and warfare at sea, where these technologies play a part in the balance of power. Much research and development occurs within the defence industry, where instincts are not broadly collaborative.

One of the first international maritime initiatives had a clear navigational agenda, for its aim was to establish an agreed prime meridian of longitude. The International Meridian Conference was held in Washington in 1884 with representatives from twenty-five countries, and decided that the prime meridian would pass through the transit circle (the principal reference instrument) of the Royal Observatory, Greenwich. The model of establishing cooperative conventions through international meetings was followed on other occasions, most famously the Safety of Life at Sea (SOLAS) convention (1914), organized in the aftermath of the Titanic disaster in 1912; this convention has been periodically modified and updated ever since.

A gathering relevant to some of the technologies mentioned in this chapter was held in London in 1946, the International Meeting on Radio Aids to Marine Navigation, where twenty-three nations were represented. A number of applications of radio had been introduced or developed during World War II and might now be applicable to peacetime use; there was also a considerable amount of kit that could be redeployed. The German *Sonne* system of long-range radio beacons for navigation was adopted by the British during the war (their own equivalent, known as *Gee*, having a more limited range), and under the name *Consol* it continued as a low-cost system for much of the century. Decca, Loran, and radar also emerged from the war as technologies for general use, with the guidance of agreed specifications.

The spirit of international cooperation following the creation of the United Nations in 1945, was an opportunity to establish a permanent agency for maritime regulation. An international conference to this end in Geneva in 1949 adopted a convention establishing the Inter-Governmental Maritime Consultative Organization, which met for the first time in 1959. The name was changed to the International Maritime Organization in 1982. The IMO is a specialist agency of the United Nations, with headquarters in London and 171 member states. Its responsibilities are very broad and, as we have noted already, they include standards for qualifications in seamanship and navigation.

A great many international bodies have an official relationship with the IMO, either intergovernmental organizations (IGOs) with agreements of cooperation, or non-governmental organizations (NGOs) granted consultative status. IAIN, for example, came into the latter category in 1976. Another example is the Nautical Institute, an international body established with the aim of representing maritime professionals and promoting standards of education and performance for its members; while its brief is general, navigation is a prominent element in its activity. A third example in the context of navigation is the Paris-based International Association of Lighthouse Authorities (IALA), founded in 1957. To return briefly to education, the International Association of Maritime Universities, with consultative status, has fifty-six university members.

To choose a single example of an IGO in cooperation with the IMO, the International Hydrographic Organization—founded in 1921 as the International Hydrographic Bureau and changing its name in 1970—is one of the earliest international bodies in our subject. The production and use of charts based on hydrographic surveys had a long history of cross-national, unofficial exchange, and it was the two major players in this history, the French and British hydrographers, who proposed an international conference—which was held in London in 1919. We have seen that this was a customary prelude to a permanent international body, which duly followed, based in Monaco, in 1921. It is the source for international standards of hydrography and charting, now including those used for electronic charts.

e-Navigation

We have seen that the shape of navigation has always been a combined function of the societal and the technical. Looking forward, new ways of envisaging navigation in the medium future have been made possible by the complementary development of global technologies and international collaboration. The notion of *e-Navigation*, promoted by the IMO with the practical support of a number of its partner organizations, imagines the coordination of a range of electronic technologies into a harmonized and, so far as possible, integrated package. E-Navigation, then, is not a single system, the ultimate and complete solution; on the contrary, it seems to have been prompted, at least in part, by fears that a satellite-based system such as GPS was becoming, inappropriately, the assumed answer to all navigational needs.

Electronic charts, delivered through what are called Electronic Chart Display and Information Systems (ECDIS) are becoming standardized and regulated. They can be coordinated with Global Navigation Satellite Systems (GNSS), such as GPS, to present positions directly on the chart viewed on a screen. Other tools can be integrated, such as Automatic Radar Plotting Aids (ARPA), introduced as an adjunct to radar technology designed for avoiding collisions at sea, and Vehicle Traffic Services (VTS), for the safe management of vessels entering and leaving port. There are other relevant systems, and their multiplication created the need for a strategy to ensure that they could work together effectively. The IMO approved a Strategy Implementation Plan for e-Navigation in November 2014.

Yet for all the sophistication of the complementary components in the navigation of the future, there is no immediate prospect of old techniques being abandoned. What we have learnt from this historical introduction to navigation at sea is very relevant to understanding the present and the future. This is demonstrated by a recent issue of *The Navigator*, a magazine published by the Nautical Institute and the Royal Institute of Navigation (RIN) for marine navigators. Devoted to 'Positioning', the general message is the importance of having a range of different techniques. While GPS is a superb tool when it works, which is almost always, there is concern about vulnerabilities that may be increasing. Interference with the signal, intentional or unintentional, is possible. The techniques listed for current use, with their strengths and weaknesses, stretch back through the history of navigation.

GNSS is of course included in the range, GPS in particular, along with its integration with ECDIS; accuracy and availability are among its strengths, potential vulnerability, and the fostering of complacency its weaknesses. Radar has the advantage of being independent of external support and it too can supplement ECDIS, yet can be compromised by bad weather and be difficult to interpret. Echo sounding is useful but has limited applicability. Loran was more resilient than GNSS and, although now largely decommissioned, eLoran may be considered for the future. Inertial navigation systems have the advantage of being self-contained, but are too expensive for commercial use.

The other systems recommended in *The Navigator* issue mentioned are familiar from earlier chapters in this book. Celestial navigation is cited as the only back-up to GNSS on the high seas; good accuracy is possible and no electronics are needed, but, as ever, weather is critical, regular practice is essential, and, it is deftly explained, the method 'needs up-to-date data (paper or electronic) for processing', that is, something like the *Nautical Almanac*. 'Dead Reckoning and Estimated Position' is there also 'based upon speed and course', though this too can be automated within ECDIS; one of its strengths is 'proven traditional technique', one of its weaknesses, as ever, 'poor accuracy over long periods'. Finally 'Visual Observations', that is, techniques of coastal navigation,

are not forgotten and return our story to where it began. The
weaknesses are as they have always been: visibility should be good
and, even then, the object observed, or whose bearing is measured,
must be within the visual range and be clearly identified.

The story of human navigation at sea has taken us across
centuries, through cultures, and around continents, pursuing a
very singular narrative, but one that has engaged with many
aspects of world history, especially the technical, mathematical,
and scientific. A few elements have accompanied us throughout,
two being the relentless and uncompromising settings of sea and
sky. Another is the practical imperative at the foundation of the
entire project. Other branches of science and mathematics can
indulge in speculation, imagination, and beauty; navigation has to
deliver on its promise: 'the shortest good way, by the aptest
Direction, & in the shortest time'.

Navigation

We have seen that mathematicians could become enthralled by
what was theoretically possible, but the verdict always rested
with what could work at sea. An involved, ungainly, or tedious
procedure was not valued for its own sake—but it would trump
an unworkable scheme, however rigorous and however elegant.
It is characteristic of science in general that theory is in dialogue
with empirical regulation, and we have followed the story of
one of the earliest disciplines where mathematics had to engage
with practical and material outcomes. If science is shaped in
history, a powerful contribution to its empirical character stems
from navigation.

Glossary of terms

The Glossary is not intended as a general guide to terminology but as an aid to reading this book. It does not include terms used only in one place in the text and explained there. Neither does it explain involved techniques, such as 'bearing and distance' or 'latitude sailing', which can be understood only by reading sections of the text. These explanations, and fuller explanations of many terms listed here, can be found using the Index. Some terms have different meanings in other contexts; here I refer only to their use in navigation.

alidade: straight rule with a pair of sights

altitude: angular distance above the horizon

amplitude: maximum displacement from rest of an oscillator, such as a pendulum or balance

apparent time: time based directly on the position of the Sun with no adjustment for the variation in its progress through the year

Arab compass: form of sky compass for finding up to thirty-two directions

armillary sphere: skeletal celestial sphere constructed of rings

astrolabe (astronomer's): instrument for astronomical calculation, based on a planispheric projection of the heavens

azimuth: horizontal angular distance

backstaff: portable instrument for measuring solar altitude by the Sun's shadow

balance: oscillating component of a watch or chronometer

bearing: horizontal direction

binnacle: housing for a magnetic compass

card: visible disc set above the magnetic needle of a compass

cardinal: refers to one of the four principal compass directions

celestial Equator: great circle on the celestial sphere, 90° from the celestial poles

celestial latitude: angular distance from the ecliptic

celestial longitude: angular distance along the ecliptic, eastwards from the vernal equinox

celestial Pole: two points on the celestial sphere, north and south, about which the heavens appear to turn

celestial sphere: imaginary sphere on which are projected the apparent positions of celestial bodies with the Earth at the centre

chart: map of the coastline and sea for navigation, or of the heavens

chronometer: specialized navigational timepiece

circumpolar: refers to stars or constellations that remain above the horizon

compass rose: circular pattern indicating the points or directions of the compass

compensation balance: oscillating element in a chronometer, designed to adjust automatically to changes in temperature

cosine: trigonometrical function, being the ratio of the adjacent side over the hypotenuse for an angle in a right-angled triangle

course: path followed by a vessel at sea; specifically the angle made with the meridian

cross-staff: instrument for measuring angles, usually altitudes

culminate: refers to a heavenly body reaching the highest point of its daily motion

cycloid: curve followed by a point on the circumference of a circle rolling along a straight line

dead reckoning: finding position from a record of direction and distance

declination: angular distance from the celestial Equator, or an alternative term for magnetic variation

departure: distance travelled east or west

deviation: angular displacement of a magnetic needle from magnetic north caused by local magnetic effects, such as of iron on a ship

dip: angle made with the horizontal by a magnetic needle free to move in a vertical plane, or the angle between the horizontal plane and the visible horizon due to the elevation of the observer

double-altitude: refers to position-finding from two ex-meridian altitudes of the Sun or a star (or two stars simultaneously)

East Indiaman: ship of the East India Company

ECDIS: Electronic Chart Display and Information System

ecliptic: apparent annual path of the Sun through the celestial sphere

ephemeris: set of tables of positions of heavenly bodies

Equator: great circle on the Earth in the plane of the celestial Equator, 90° from the poles

equinoctial sundial: sundial where a shadow or spot is projected on to the plane of the Equator and the hours are marked at equal intervals

equinox: one of two points on the celestial sphere where the Sun's annual path crosses the Equator, or a time when this occurs

escape wheel: component of an escapement in a clock or watch

escapement: part of a clock or watch mechanism which controls the release of the force of the mainspring or driving weight and maintains the movement of the oscillator

ex-meridian: refers to celestial bodies not on the observer's meridian

fathom: unit of depth measurement, based originally on the distance between outstretched hands

gimbal mount: ring with pairs of inner and outer supporting pivots at right angles, arranged to maintain a compass or chronometer level

GNSS: Global Navigation Satellite System

GPS: Global Positioning System

great circle: circle on the surface of a sphere in a plane passing through the centre of the sphere

Guard Stars (or Guards): two stars in the constellation of Ursa Minor

Gunter's rule: instrument with logarithmic scales used in navigational calculations, especially in Mercator sailing

hemisphere: half of a sphere

horizon: line where the sea appears to meet the sky, if the view is not obscured

horizon glass: fixed, half-silvered mirror of an octant, sextant, or reflecting circle

hydrography: discipline of surveying at sea and on the coast, and the making of charts

hyperbolic: in the form of the conic section known as the hyperbola

IMO: International Maritime Organisation

index mirror: moving mirror of the octant, sextant, or reflecting circle

isochronal: describes an oscillation whose period is independent of amplitude

kamāl: form of altitude measuring instrument

knot: unit of speed, one nautical mile per hour

latitude: angular distance of a position on Earth from the Equator

lead and line (or sounding lead): instrument for measuring depth

leeway: lateral drift of a ship in the direction away from the wind

log: instrument for measuring the speed of a ship, or a navigational journal kept on a voyage

longitude: angular distance on Earth, measured parallel to the Equator from a chosen meridian

loxodrome: line on a sphere cutting all the meridians at the same angle

lubber-line: mark on the compass or binnacle to indicate the direction of the bow of the ship

lunar distance: angular distance from the Moon to another heavenly body, or a method of finding longitude based on such a measurement

lunar mansions: division of the zodiac used in Arab or Islamic astronomy

lunars: method of finding longitude by lunar distance

magnetic compass: instrument indicating direction, using the north-seeking properties of a magnet free to turn in a horizontal plane

magnetic variation: local angle between true north and the direction taken by a magnetic needle (magnetic north)

mariner's astrolabe: circular brass instrument for measuring celestial altitude

mean time: time by the calculated motion of the mean Sun, which averages out the annual cycle of variation in the apparent motion of the Sun, and is the time shown by an ordinary clock

meridian: great circle on the celestial or terrestrial sphere passing through the poles, or the plane containing that circle; without further qualification, it refers to the meridian of the observer

meridional parts: specified variable in the Mercator projection, being the angles along a meridian between the Equator and parallels of latitude, expressed in minutes of arc at the Equator

Nautical Almanac: annual volume of astronomical tables and other information for navigation, specifically here the annual ephemeris of Great Britain

nautical mile: measure of distance at sea formerly intended as the length covering a minute of arc along a great circle; now fixed at 1,852 metres

nocturnal: instrument for finding time from the orientation of the sky around the celestial Pole

noon sight: operation of taking the altitude of the Sun at midday

octant: 45° instrument for measuring angles up to 90° between distant bodies, using reflection to bring images into coincidence

parallax: difference in an observed position or direction due to a difference in the viewpoint of the observer(s)

parallel (parallel of latitude): latitude line

phase: particular point in a periodic phenomenon or wave

phase difference: the difference between the same phase of two waves with the same frequency, measured as an angle or a time

pilotage: navigation by observation of the coast and sounding depth

Pleiades: cluster of stars in the constellation of Taurus

polar distance: angular distance from the celestial Pole

Polaris: brightest star in the constellation of Ursa Minor, close to the north celestial Pole, also called the Pole Star or the North Star

portolan chart: late-medieval and early-modern style of navigational chart, centred on the Mediterranean and crossed by radiating rhumb lines

position line: line on a chart which passes through a ship's position at some point

projection: geometrical technique for representing a figure or set of positions on a surface, here especially of a spherical surface on a plane one

PZX triangle: figure on the surface of the celestial sphere whose three sides are arcs of great circles joining the Pole, the observer's zenith, and the position of a heavenly body being observed

quadrant: instrument for measuring angles between distant bodies with a graduated arc of 90°

rate: known and regular discrepancy between the progress of a watch or chronometer and the true passage of time

reflecting circle: full-circle instrument for measuring angles between distant bodies, using reflection to bring images into coincidence

refraction: deflection of light from a straight path as it passes through different densities of medium, here specifically light from a heavenly body passing through the Earth's atmosphere

repeating circle: instrument where a sequence of measurements of an angle can be accumulated and averaged

rhumb: compass bearing, or the course of a ship following a constant bearing

right ascension: angular distance, expressed in units of time, along the celestial Equator eastwards from the vernal equinox

run: progressive change in the position of a ship

sea astrolabe: see 'mariner's astrolabe'

secant: trigonometrical function, being the ratio of the hypotenuse over the adjacent side for an angle in a right-angled triangle, which is the reciprocal of the cosine

sector: a folding rule with pairs of lines used for calculation

sextant: 60° instrument for measuring angles up to 120° (and often more) between distant bodies, using reflection to bring images into coincidence

sky compass: technique of finding direction from the positions of rising or setting stars

solstice: time when the Sun is at its maximum distance from the celestial Equator, happening twice a year

sounder: instrument for measuring depth

Southern Cross: group of stars in the constellation 'Crux' in the southern celestial hemisphere

spherical triangle: figure on a spherical surface whose three sides are arcs of great circles

starboard: side of a ship to the right when facing forward

tangent: trigonometrical function, being the ratio of the opposite over the adjacent side for an angle in a right-angled triangle

transom: adjustable cross-piece of a cross-staff (see also 'vane')

traverse: path taken by a ship, which may comprise a number of different bearings

traverse board: an instrument for recording the bearings steered in the course of a watch, or an instrument for calculating the outcome of a set of bearings and distances sailed

Ursa Minor: constellation in the northern celestial hemisphere, the Little Bear

Ursa Major: constellation in the northern celestial hemisphere, the Great Bear, contains 'the Plough' or 'Big Dipper'

vane: adjustable cross-piece of a cross-staff (see also 'transom'), or the sights of a mariner's astrolabe or of a backstaff

variation compass: an instrument for measuring magnetic variation

watch: as well as a portable timepiece, one of the divisions (usually into four hours) of the nautical day (from noon to noon) used to regulate the duties of the crew

way: course covered by a ship

wind rose: circular compass rose, usually on a chart, with the directions marked by names or initials of the winds of the Mediterranean

winds: nomenclature for direction formalized from traditional names for winds in the Mediterranean

zenith: point on the celestial sphere directly above the observer

zenith distance: angular distance from the zenith

zodiac: band of constellations close to the ecliptic, through which the Sun passes on its annual apparent path

Further reading

Among general works on maritime history with significant sections on navigation, John B. Hattendorf (ed.), *The Oxford Encyclopedia of Maritime History*, 4 vols (Oxford: Oxford University Press, 2007) is comprehensive and up to date. For something at a more manageable size, there is Donald S. Johnson and Juha Nurminen, *The History of Seafaring: Navigating the World's Oceans* (London: Conway Maritime, 2007); or Lincoln P. Paine, *The Sea and Civilization: A Maritime History of the World* (London: Atlantic Books, 2014).

General histories of navigation must still include Eva G.R. Taylor, *The Haven-Finding Art: A History of Navigation from Odysseus to Captain Cook* (London: Hollis and Carter, 1971), though first published in 1956; and David W. Waters, *The Art of Navigation in England in Elizabethan and Early Stuart Times* (London: Hollis and Carter, 1958). Also useful are J.B. Hewson, *A History of the Practice of Navigation* (Glasgow: Brown, Son and Ferguson, 1983), which is good on electronic navigation prior to satellites (he writes when GPS was under construction, 'Hopefully this incredible accuracy will be available to all mariners.'); and W.E. May, *A History of Marine Navigation* (Henley-on-Thames: Foulis, 1973) (with a valuable chapter on recent developments at the time of writing). The latter two titles have the advantage of being written by men with experience of the sea. J.E.D. Williams, *From Sails to Satellites: The Origin and Development of Navigational Science* (Oxford: Oxford University Press, 1992) offers a rather individual account with a number of particular strengths. A valuable recent work is Mark Denny, *The Science of Navigation: From Dead Reckoning to GPS* (Baltimore: Johns Hopkins, 2012).

Written in the context of a more personal narrative is David Barrie, *Sextant: A Voyage Guided by the Stars and the Men who Mapped the Ocean* (London: William Collins, 2014).

The recent interest in the longitude problem originated through a symposium held at Harvard University in 1993, the proceedings being published as, William J.H. Andrews, *The Quest for Longitude* (Cambridge, MA: Harvard University, 1996). The extraordinarily popular book, Dava Sobel, *Longitude: the True Story of a Lone Genius who Solved the Greatest Scientific Problem of his Time* (London: Fourth Estate, 1995) has the disadvantage of being very one-sided, despite the more scrupulous work found in earlier books, such as Rupert T. Gould, *The Marine Chronometer: Its History and Development* (London: Holland Press, 1960); and Humphrey Quill, *John Harrison: The Man who Found Longitude* (London: John Baker, 1966). It is easier to recommend Derek Howse, *Greenwich Time and the Discovery of Longitude* (Oxford: Oxford University Press, 1980); and Derek Howse, *Nevil Maskelyne: The Seaman's Astronomer* (Cambridge: Cambridge University Press, 1989). A welcome recent corrective is Richard Dunn and Rebekah Higgitt, *Ships, Clocks and Stars: The Quest for Longitude* (Collins: Glasgow, 2014).

Recommended accounts of particular topics follow; in each case the title is sufficiently clear to indicate the subject and its relevance to this book.

James A. Bennett, *The Divided Circle: A History of Instruments for Astronomy, Navigation and Surveying* (Oxford: Phaidon Christie's, 1987)

Tony Campbell, 'Portolan Charts from the Late Thirteenth Century to 1500', in J.B. Harley and David Woodward (eds), *The History of Cartography*, vol. 1 (Chicago: University of Chicago Press, 1987), pp. 371–463

Anthony R. Constable and William Facey, *The Principles of Arab Navigation* (London: Arabian Publishing, 2013)

Charles H. Cotter, *A History of Nautical Astronomy* (London: Hollis and Carter, 1968)

Barry Cunliffe, *The Extraordinary Voyage of Pytheas the Greek* (London: Allen Lane, 2001)

David Fabre, *Seafaring in Ancient Egypt* (London: Periplus, 2004/5)

Louise Levathes, *When China Ruled the Seas: The Treasure Fleet of the Dragon Throne, 1405-1433* (Oxford: Oxford University Press, 1996)

David Lewis, *We, the Navigators: The Ancient Art of Landfinding in the Pacific* (Honolulu: University of Hawaii Press, 1972)

Geoffrey J. Marcus, *The Conquest of the North Atlantic* (Woodbridge: Boydell, 2007)

Fik Meijer, *A History of Seafaring in the Classical World* (London: Croom Helm, 1986)

Willem F.J. Mörzer Bruyns, *The Cross-Staff: History and Development of a Navigational Instrument* (Amsterdam, Scheepvaart Museum, 1994)

Joseph Needham, ed. Colin A. Ronan, *The Shorter Science and Civilisation in China*, vol. 3 (Cambridge: Cambridge University Press, 1986)

Patrice Pomey (ed.), *La navigation dans l'antiquité* (Aix-en-Provence: Edisud, 1997)

G.S. Ritchie, *The Admiralty Chart: British Naval Hydrography in the Nineteenth Century* (London: Hollis and Carter, 1967)

Alan Stimson, *The Mariner's Astrolabe: A Survey of Known, Surviving Sea Astrolabes* (Utrecht: Hes, 1988)

Peter Whitfield, *The Charting of the Oceans: Ten Centuries of Maritime Maps* (London: British Library, 1996)

Index

Navigation

SOCIAL MEDIA
Very Short Introduction

Join our community
www.oup.com/vsi

- Join us online at the official Very Short Introductions **Facebook** page.
- Access the thoughts and musings of our authors with our online **blog**.
- Sign up for our monthly **e-newsletter** to receive information on all new titles publishing that month.
- Browse the full range of Very Short Introductions online.
- Read **extracts** from the Introductions for free.
- If you are a teacher or lecturer you can order inspection copies quickly and simply via our website.

ONLINE CATALOGUE
A Very Short Introduction

Our online catalogue is designed to make it easy to find your ideal Very Short Introduction. View the entire collection by subject area, watch author videos, read sample chapters, and download reading guides.

http://fds.oup.com/www.oup.co.uk/general/vsi/index.html